THE PHYSICIAN AS CAPTAIN OF THE SHIP: A CRITICAL REAPPRAISAL

PHILOSOPHY AND MEDICINE

Editors:

H. TRISTRAM ENGELHARDT, JR.

Center for Ethics, Medicine, and Public Issues, Baylor College of Medicine, Houston, Texas, U.S.A.

STUART F. SPICKER

School of Medicine, University of Connecticut Health Center, Farmington, Connecticut, U.S.A.

VOLUME 29

THE PHYSICIAN AS CAPTAIN OF THE SHIP: A CRITICAL REAPPRAISAL

Edited by

NANCY M. P. KING, LARRY R. CHURCHILL,
and ALAN W. CROSS

*University of North Carolina School of Medicine,
Chapel Hill, North Carolina, U.S.A.*

SPRINGER-SCIENCE+BUSINESS MEDIA, B.V.

Library of Congress Cataloging-in-Publication Data

CIP

The Physician as captain of the ship : a critical reappraisal / edited
by Nancy M. P. King, Larry R. Churchill, and Alan W. Cross.
p. cm. — (Philosophy and medicine; v. 29)
Papers presented at a symposium, University of North Carolina at
Chapel Hill, May 15–17, 1986.
Bibliography: p.
Includes index.
ISBN 978-94-017-3736-4 ISBN 978-0-585-27589-5 (eBook)
DOI 10.1007/978-0-585-27589-5
1. Physician and patient—Congresses. 2. Authority—Congresses.
3. Medicine—Philosophy—Congresses. I. King, Nancy M. P.
II. Churchill, Larry R., 1945- . III. Cross, Alan W., 1944-
IV. University of North Carolina at Chapel Hill. V. Series.
R727.3.P478 1987
362.1—dc19 87–30771
CIP

Originally published by D. Reidel Publishing Company, Dordrecht, Holland in 1988
Softcover reprint of the hardcover 1st edition 1988

FOR RAY DUFF

TABLE OF CONTENTS

ACKNOWLEDGMENTS

The articles collected in this volume began as papers presented at a symposium, entitled "The Physician as Captain of the Ship: A Critical Reappraisal," that was held at the University of North Carolina in Chapel Hill on May 15–17, 1986. We have a great many sponsors and supporters to thank. First and foremost among our sponsors is the Medicine and Society Program of the North Carolina Humanities Committee and the Duke Endowment; many thanks are also due to the American Medical Association Education and Research Foundation and the Burroughs Wellcome Corporation. We are also most grateful for the support of the UNC School of Medicine: Dean Stuart Bondurant and Glenn Wilson, Chair of the Department of Social and Administrative Medicine, were both instrumental in the success of the symposium from which this volume has grown. The editors of this Series, Stuart F. Spicker and H. Tristram Engelhardt, Jr., freely gave priceless assistance and advice.

A special debt is owed to those who made a program as large and complex as ours run as well as it did. The staff of the School of Medicine's Office of Continuing Education, and especially Betty Neilson, were a constant source of help. Ruth Knight of North Carolina Memorial Hospital's fiscal division skillfully managed our funds. Staff of the Hanes Art Center and the Ackland Art Museum helped make the program as pleasant as it was productive. And finally, the secretarial staff of the Department of Social and Administrative Medicine deserve endless praise for their tireless efforts. Nothing could have happened without them.

Our greatest thanks are reserved for those participating faculty whose significant contributions to the symposium are not directly reflected by publications in this volume: Joe Graedon, MPharm; Martha Henderson, MSN, MDiv; Elizabeth Minnick, PhD; and Ruel W. Tyson, BD. And we are grateful, finally, to our audience, for embracing our metaphorical vision and sailing with it – and with us the faculty – to the farthest shores.

NANCY M. P. KING
LARRY R. CHURCHILL
ALAN W. CROSS

INTRODUCTION

> "The fixed person for fixed duties, who in older societies was such a godsend, in the future will be a public danger."
>
> ALFRED NORTH WHITEHEAD

Twenty years ago, a single legal metaphor accurately captured the role that American society accorded to physicians. The physician was "captain of the ship." Physicians were in charge of the clinic, the operating room, and the health care team, responsible – and held accountable – for all that happened within the scope of their supervision. This grant of responsibility carried with it a corresponding grant of authority; like the ship's captain, the physician was answerable to no one regarding the practice of his art.

However compelling the metaphor, few would disagree that the mandate accorded to the medical profession by society is changing. As a result of pressures from a number of diverse directions – including technological advances, the development of new health professionals, changes in health care financing and delivery, the recent emphasis on consumer choice and patients' rights – what our society expects physicians to do and to be is different now. The purpose of this volume is to examine and evaluate the conceptual foundations and the moral implications of that difference.

Each of the twelve essays of this volume assesses the current and future validity of the "captain of the ship" metaphor from a different perspective. The essays are grouped into four sections.

In Section I, Russell Maulitz explores the physician's role historically. He identifies three types of authority: professional, which is concerned with boundary control and the internal pecking order; cultural, which concerns physician-patient expectations and behaviors; and legal, which is a distillation of the other two. His chief aim is to show how physicians have held on to their authority, and his claim is that this has been

Nancy M. P. King, Larry R. Churchill, and Alan W. Cross (eds.)
The Physician as Captain of the Ship: A Critical Reappraisal, xi–xv.

achieved mainly through the power to name and diagnose. Currently, Maulitz claims, the physician has new partners in naming – the economist, the manager, the policy scientist. These new partnerships, combined with the inherent plasticity and contingency of disease concepts, may well lead to a diffusion of responsibility.

Joan Lynaugh also looks at authority in an historical context, but her essay investigates the distribution of work and authority between nursing and medicine. The rise of nursing as a distinctive profession, she believes, coincided roughly with the use of hospitals as the preferred place to be ill. "Captain" rhetoric intensified as doctors became more dependent on hospitals as arenas for giving care. Historically, nurses sought to distinguish between physicians' authority over patient care and their authority over nurses themselves. More often, Lynaugh notes, this distinction failed to be made and personal loyalty to the physician was demanded.

Lynaugh also explores divergence in the authority assumed by nurses and physicians through a look at language differences. For example, what for nurses is a "home visit," for doctors is a "house call." What nurses call a "care plan," doctors refer to as a "regimen." Lynaugh believes these differences are more than nominal. Her paper ends with a call for greater parity between nurses and physicians as a safer way to care for patients.

A legal perspective on the captaincy question is provided by Judith Areen in her essay, "Legal Intrusions on Physician Independence." She claims that although many intrusions are technically legal, the law itself is not an intruder, but a stalking horse. The law is not the primary cause of the diminished status of doctors, and the real villains are, ironically, offspring of medicine's great success. Three causes of diminished physician authority are discussed in detail: (1) increased expectations of medicine from the general public; (2) greater awareness of the limitations of medicine; and (3) a push for greater efficiency and accountability for costs. All cause tension within the physician role.

Areen does see hopeful signs: reforms in the malpractice system and the trend toward reliance on families as proper surrogates in end-of-life choices. The greatest challenge, she claims, is whether physicians can retain their standards as advocates of efficiency seek to steer the ship.

H. Tristram Engelhardt, in response to Maulitz, Lynaugh, and Areen, lists eight issues he finds explicitly or implicitly in all three papers. He clarifies these issues through an examination of the multiple senses of 'authority', notably the differences between being *in* authority and being *an* authority. Engelhardt reminds us that whatever sort of authority we wish to vest physicians with, we should not assume that any particular sort is written in stone, and we cannot enjoy the fruits of all of them at once. We should choose carefully just what sort of physician-captain we want.

The three papers in Section II explore a more practical and interpersonal dimension of the physician's role. In "Team Medicine in the NICU: Ship or Flotilla of Lifeboats?" neonatologist Ernest Kraybill seeks to dispel some of the majestic imagery surrounding care of handicapped newborns. He explores in brief the history of neonatology and the forces that combine to make "team medicine" a necessity in this setting. His essay is a description of how teams function within a commons fraught with uncertainty, with special attention to nurses, respiratory therapists, nutritionists, physical therapists, social workers and house staff. Although authority is clearly shared, Kraybill believes it is still appropriate for physicians to chart the course, though he feels "captain" may be less appropriate than "team leader."

David Barnard, by contrast, presents what he terms "patient-centered perspectives on the health care team," which he claims are neglected in the idioms of "captains," "mates," and other nautical metaphors. Barnard's essay is a sweeping indictment of "medicine's selective inattention to the impact of illness and treatment on the patient's daily life, psychological state and personal values." His paper is centered on the analysis of illness as a social event, that is, as an instance of human connectedness. The analysis leads the reader through John Bowlby's work on "attachment behavior" and D. W. Winnicott's work on "the holding environment." Against the conventional wisdom, Barnard claims that these authors allow us to see the regression needs of the sick and the development of their autonomy as linked and interdependent – "genuine autonomy can only emerge from a matrix of dependence." Finally, he lists criteria for a patient-centered notion of health care.

In response to both Kraybill and Barnard, Peter Morris addresses the captaincy motif through his experiences as a practitioner in the coal country of Appalachia. His experiences as a provider on equal footing with nurses in the Kentucky Frontier Nursing Service lead him to stress the degree to which all professional and patient roles are learned in a social context. He contrasts vertical with more egalitarian team structures and suggests that if physicians seek to be technocratic elite they will find their authority eroded.

Section III considers changes in the physician's role due to technological and economic innovations. Robert Cook-Deegan's contribution to this volume, "The Physician and Technological Change," investigates the way in which rapid technological change has led to mistrust of physicians' moral authority. His essay is primarily concerned with the secondary, or indirect, effects of the expansion of technology, but includes a discussion of how medical technologies are created, developed, and disseminated, and how they eventually pass into obsolescence. As medical care becomes more technologically driven, Cook-Deegan believes that the conflict in the physician's role between altruism and self-interest (and between strictly medical and new fiscal duties) will be intensified.

While Cook-Deegan assesses the impact of technological change, Stuart Spicker focuses on economic aspects. His essay, "Marketing Health Care: Ethical Challenge to Physicians," questions the usual assumption of an inherent conflict between the norms of business and medical ethics. Spicker contends that the profit motive has always existed, though he believes new mechanisms may be required to control what he terms "excessive self-interest" which may accompany current pro-competition and for-profit initiatives. Spicker also speculates that traditional intraprofessional loyalties will be tested by the economic canons that regulate many practices.

Wendy Mariner, in a probing and synthetic essay, weaves together and recasts the basic themes of Cook-Deegan and Spicker. She posits three generally acceptable goals of medicine – efficiency, effectiveness, and equity of access – and argues that we have failed as a society to decide how we want to balance them, or, more generally, to specify a purpose for our health care system. The result is that we try each of these goals seriatim, having moved most recently from equity to efficiency. Mariner also takes issue with Spicker for what she takes to be his reliance on the good will of managers to maintain equity and effectiveness in a climate that rewards efficiency and profit. The solution, Mariner thinks, is to decide first what purposes our system will serve, and then enact legislation to steer in that direction.

The last two papers in this volume, under Section IV, deal with ethics committees and their impact on the physician's authority. Raymond Duff approaches his assessment of committees through a more generic consideration of the human phenomena of helplessness and healing. Medicine's paternalistic tradition, Duff claims, combined with the modern advent of technical-biological expertise, paved the way for the physician to become "captain of the ship." Yet this role too often increases rather than alleviates the patient's helplessness and is ultimately counterproductive for healing. The appropriate models for healing actions are Cabot and Codman, rather than Osler, says Duff.

Regarding ethics committees Duff is skeptical. He worries that the traditions of medical paternalism may simply be transferred from individual physicians to their bureaucratic equivalents in committees and that patients and their values will remain unheeded.

Nancy King, in an essay entitled "Ethics Committees: Talking the Captain Through Troubled Waters," takes a different approach. She argues that the authority of physicians originally comes from society, and was never an aboriginal possession of the profession. Thus, critical appraisal of the physician as captain is fitting and less a revolution than a recognition of a proper relationship. In keeping with this line of thinking, King compares the work of ethics committees with two other sorts of hospital committees, viz., morbidity and mortality conferences and institutional review boards, with the function of ethics committees falling somewhere roughly between the other two. Then, selecting a favorite phrase of Raymond Duff, King considers the extent to which ethics committees can be "moral communities." The proper role of the ethics-committee-as-moral-community is not adversarial adjudication based on rights, but conciliation and transformation in dealing with issues where rights are unclear. The ultimate purpose, King proposes, is to show that "our own moral sensibilities . . . can be exercised responsibly *when certainty eludes us*."

The essays in this volume do not provide a map for the future of medicine. They do, however, raise and discuss essential questions about the relationships between physicians and society, and medicine as a profession and the larger culture. We ignore them at our peril.

LARRY R. CHURCHILL

SECTION I

THE CAPTAIN'S AUTHORITY: SOURCES AND SCOPE

RUSSELL C. MAULITZ

THE PHYSICIAN AND AUTHORITY: A HISTORICAL APPRAISAL[1]

I. INTRODUCTION

As far as medicine and health care are concerned, most of us probably share the feeling that the ground is shifting under our feet. Things are changing. The litany is everywhere: for-profit hospitals, the senescence of the biomedical model, Diagnosis Related Groups, the malpractice crisis, and so on. How did we get to this pass?

I can begin, perhaps, by looking at some of the features of the change going on around us in medicine, and then discuss some of the reasons. I hope first briefly to characterize the changing image of the physician in American society. Next I will comment on the elements that combine to support his (or her) basis of authority and power. Then I will advance some historical examples. I will not dwell on the historical antecedents of modern physicians, or those earlier practitioners' claims to a certain social or legal status *per se*. I will try rather to examine the *contingent* nature of those men's claims.

I will concentrate on how both professional and cultural success has been *negotiated*. Success has rarely befallen groups of doctors for their good looks or high social status, at least not in recent decades. After all, the captain of the ship has to be able to show that expertise in navigation is more important than, say, knowledge of championship checkers or show-jumping. He or she must then demonstrate that navigation is actually *important*. This point is actually a subtle one, in part due precisely to the fact that it seems so self-evident. (In medicine, though, the navigation charts get redrawn to an almost unrecognizable degree every time a major new theory or technique arrives on the scene.) I will describe several instances in which medical men shored up their professional and cultural authority. Then I will be in a position to talk about whether and how some of those negotiations may now be breaking down. I conclude with a glance at the physician's place in America at the end of the twentieth century.

Nancy M. P. King, Larry R. Churchill, and Alan W. Cross (eds.)
The Physician as Captain of the Ship: A Critical Reappraisal, 3–21.

II. FORMS OF AUTHORITY

Pictures of physicians as beneficent authority figures or conniving manipulators have a signal virtue which is also a significant flaw. They condense several layers of meaning into a single, seemingly impregnable image. It is worth unpacking these images, especially in connection with the medical man's authority. Why do we follow the captain-physicians? Is it because of power we ascribe to them, given their conspicuously superior expertise? Is it because of power they have seized and brainwashed us into thinking they always had? Or is it because of power we have to allocate somewhere, and this figure just happens to be in the right place at the right, secularized time? When we look at the physician and his or her captaincy of the health care enterprise, then, it might help structure the argument to look at the different potential forms of authority arrogated to the modern M.D.

In this simplified but still useful version of a triangular set of relationships currently popular among social scientists, some will detect a resemblance to the early part of the argument advanced by Paul Starr in his recent, much-ballyhooed study of *The Social Transformation of American Medicine* ([36], pp. 9–29). (The resemblance is deliberate as I try to structure my own argument against the backdrop of recent discourse on the reinvention of American medical care. But I share other recent critics' profound reservations about the arguments Starr develops to link the elements he so adroitly displays in this ingenious work; e.g., [34].)

The doctor's *legal* authority (and concomitant responsibility) is in some sense the distillation of all the other ideological and societal forces at work. I still have to sign prescriptions written by nurse-practitioners on the team to which I belong, and in Pennsylvania I am still legally responsible for any adverse effects. Legal perspectives form one of the leitmotifs of any conference on the captaincy of the physician, but I will pass over it after a brief comment. There are various dimensions to it, including the juridical, discussed elsewhere in this volume by Dr.

Areen; the ethical, which Drs. King and Duff discuss; and the linguistic or ideological, which Dr. Theodore Brown from Rochester explored just a few years ago in his essay for another conference in this series [4]. (He made the still valid point, upon which I hope to build, that "bioethical discussion . . . must shift its focus away from individual relationships to complex, interactive contemporary realities and the incompletely explored philosophical problems of 'authority' and 'responsibility'. . . .")

Instead of pursuing a legal history of the physician's vaunted position at the helm, I want to look at its twin buttresses, as it were, in the two key sets of justifications for that position. If one asks what were the dynamics of this position, what *drove* the medical profession's legal justification of its unique hand on the tiller, the answers precipitate around the two kinds of authority depicted above. Professional authority, with which I deal first, concerns the physician's efforts to draw boundaries. How big should the profession be allowed to grow? Who gets in? Who is kept out? What is the pecking order? Internal as well as external boundaries count, as I will show momentarily.

In the end, though, professional authority relies on matters like solidarity. By professional authority I wish to denote boundary maintenance by providers and, especially, educators, while saying little about what goes on *within* the boundaries. Cultural authority, on the other hand, denotes that content.[2]

III. HIERARCHY AND SOLIDARITY

I turn now to the ways in which the physician has held on to the captaincy. One must first grasp the notion of hierarchy, an idea that spans all the humanistic and social science disciplines represented in this volume, though it sometimes holds rather different meanings for historians, philosophers, and sociologists.

Is the physician still at the pinnacle of the "medical beehive," which happens to be the bridge of our metaphorical ship? If so, how has that uppermost position been maintained? What are the dynamics of the competitive process by which doctors, nurses, midwives, charlatans, pharmacists, homeopaths, and the rest have created niches for themselves? That competition itself is a "motor" of change that will be apparent to all of us. To the extent that the teamwork concept developed in the post-World War II era [4], it was in part because cooperation between different kinds of health care providers was one response to the specter of bruising competition. Cooperation and competition always coexist in a sort of shifting equilibrium: if you cannot beat them, join them.

In thinking about how health professionals retain their niches and maintain boundaries – in modern parlance, their "market position" – it is important to recognize one other feature of hierarchical behavior: it is rampant *within* medicine as well as without. Within most professions there is some sort of a pecking order, depending on matters like seniority and institutional location. Within medicine, there is a pecking order of specialties, an order that has been far from invariant over time, as George Rosen began to illustrate two generations ago [32]. Even within seemingly homogeneous specialties like internal medicine [37], individuals and subspecialties are arrayed hierarchically. In less highfalutin terms, we all know that medicine has both its QE IIs and its tugboats. Their captains may belong to the same club, but they do not necessarily sit down at the same table.

It is important to consider the issue of this hierarchical and often competitive state of affairs *within* medicine. If we want to understand the forces that shift the equilibrium between cooperation and competition – hence between physician as imperious captain and physician as team member – this is a critical place to cast our gaze. To test this idea, one may examine a few vignettes in which physicians saw their own place in the hierarchy in some way threatened.

Consider first the following scenario. The members of a cadre of university-trained physicians are beginning to increase dramatically in number. Their services are becoming considerably more abundant than elsewhere, one result being redistribution of those services: they begin moving from major urban centers into smaller towns and rural areas. Practitioner-to-patient ratios are changing rapidly. One exasperated doctor is heard to lament that the

> number of physicians has increased so vastly that not only are the cities filled with them, but every single little borough, every little village, every little hamlet can have its community physician at little cost. As the incomes of these doctors are inadequate, for their living they must resort to expedients which disgrace the art. Even in the cities where the physicians practice with great decorum, there are now doctors who degrade the practice through subservience, sycophancy, and buffoonery (quoted [8], p. 86).

What is the aspiring young member of the medical elite to do in this situation? Reach out – reach down, as it were – to new markets, even if it means finding one's new patients in unaccustomed economic and social strata. The pre-paid provider, often paid from the public purse, becomes a more commonplace fixture on the medical scene.

In some of its features this scenario might be recognizable to today's health planner, but it describes northern Italy at the beginning of the seventeenth century. Medical education was the engine, fueled by the new anatomy and physiology of Andreas Vesalius and his heirs. Pisa, Florence, Padua, and Milan became magnets attracting ever-increasing numbers of students, not only from the Italian states but from as far away as England [6]. A physician surplus ensued (or, more properly, the *perception* of a physician surplus). For a while public expenditures in a thriving late Renaissance Italy kept pace. The elite physician remained at the helm, presumably at the expense of other, separate providers, such as midwives or barber-surgeons, but reaching outward and downward to an expanded clientele.

Now consider another scenario. By negotiating new markets and enlarging his (still exclusively male) domain, the physician had retained his place as captain of the ship. But lately his craft had been shipping water rather badly. Radical changes had swept the length and breadth of society. With a mixture of horror and approbation the profession watched the abolition of old style privilege, a civil rights revolt, and an enervating war that was beginning to wind down of its own adventurous weight. Mainstream medicine had already reached out to certain lower order doctors who seemed near enough to the mainstream, and with them fashioned a sort of *rapprochement*. But, largely as a result of the recent wars, large numbers of secondary practitioners, trained in the trenches as it were, were mustering out of military service and demanding a place in the professional sun.

This was not the story of post-Viet Nam America, but of post-revolutionary France under Napoleon, a France which from 1803 on boasted a subclass of *officiers de santé* or health officers. Since 1794 the physician elite had already made their peace with the surgeons, now "lower order" in the perceptions of few but their traditional physician rivals. The surgeons in turn offered wider markets and, as a sweetener, a fair amount of choice Paris real estate [1, 12, 40]. The *officiers*, on the other hand, remained a thorn in the side of the newly unified physicians and surgeons alike, all the while justifying their continued role by their willingness to practice in poorer, undermedicalized departments. For their part, the physicians and surgeons exerted professional solidarity and authority by staffing the medical juries set up to examine and regulate the *officiers* [22].

Only in 1892 was the role and title of the *officier de santé* finally, fully abolished [14]. But he had already long since ceased to be a real threat, outnumbered as he was almost six to one by medical doctors. Why did the mainstream physician continue to feel threatened? I offer but a hypothesis. Another crisis was under way, this time within the medical profession *per se*. During the decade and a half between 1876 and 1891, in the face of a population that (uniquely in nineteenth century Europe) had grown essentially not at all, the medical profession increased its numbers disproportionately, probably on the order of 10 per cent absolute growth. Can the dissolution of the *officier* cadre have been an exercise in "nibbling at the edges"? By the end of the century the elite physician could hardly have considered a radical amputation, say, of the surgeon class – its members were, after all, now part of the mainstream. This state of affairs led one leading spokesman to sound the alarm to the students of the Paris medical faculty:

> I will not disguise from you the fact that the present situation is grave and, far from improving, seems to worsen daily. For, while the number of physicians is increasing, that of the sick . . . shows a tendency, thanks to the progress of private and public hygiene, to diminish in very large proportions.
>
> You know, gentlemen, that the physician does not make a fortune, and I fear greatly that if we do not check the trend that is drawing so many young people toward the study of medicine, we will see the number of the deprived in the medical profession grow by enormous proportions ([3], p. 26).

When I first read these words half a dozen years or so ago, I thought about America in the early 1980s. Medical schools were just then approaching their largest medical school graduating classes in history, a figure that went on to crest about two years ago. It occurred to me then we would likely see the captains of some of our largest medical ships coming out with proposals of their own to trim away, as it were, at the edges. This, as you know, has since come to pass. We are now witnessing heated debates both about whether to curb training and subvention of U.S.-based foreign-born foreign medical graduates, and whether to curb the size of U.S. medical school classes [23, 29, 39].

Consider now one final scenario, this one from twentieth century America. Medicine has become increasingly specialized. A more fully urbanized and well-educated population has raised its expectations of the doctor's expertise. But again, on top of this, the medical profession has grown even faster than the general population (at least 25% as opposed to ca. 20% [41]). The president of the American Medical Association (AMA) strode to the podium at its national meeting:

> It has been estimated that there is an average of one physician to six hundred of the population of the United States at the present time. The natural increase in the population of the country and the deaths in the rank of the profession, make room each year for about 3,000 physicians, based on the proportion of one physician to six hundred of the population. With 5,000 or more graduates each year, a surplus of ten thousand physicians is thrown on the profession, overcrowding it, and steadily reducing the opportunity of those already in the profession to acquire a livelihood. The evil of an overcrowded profession is a sufficient cause of complaint, but the cause thereof is the important point for us to consider and, if possible, remove. To correct the evil, the ease and facility with which a medical degree may be secured in this country must be diminished (Frank Billings at the 1903 Association meeting [2]).

Perhaps the only thing in Frank Billings' lamentation that gives it away as a product of the first decade of this century, and not the 1980s, is the number of graduates. His five thousand has tripled and then some, but the sentiments remain very much the same. There is a sense of erosion in the physician's authority, now as then, with the loss of autonomy potentiated (in the perception and in the reality) by the specter of physician unemployment – by the loss of political and professional bargaining power.

Responses to this state of affairs range from the shrill and probably ineffectual to the considered and at least potentially effective. In the period just before World War I the response was, to oversimplify somewhat, one of narrowing supply rather than (as in the earlier examples) expanding demand. Physician educators, especially full time clinical scientists, with the aid of the reformist AMA and Abraham Flexner, engineered a remarkable slow-motion collapse – call it a controlled implosion – of the medical educational establishment. The idea of an elite medicine was well served, the physician's professional and cultural authority securely shored up for the remainder (or most of the remainder) of the twentieth century.

The successful new disciplines of biochemistry, pathology, and bacteriology made it possible for mainstream medicine, by embracing the new science, to argue convincingly (and seemingly without self-interest) for what amounted to a restraint on the supply of physicians. What was important in the process was the presence of two ingredients: a perception of overmedicalization, creating the brute motive force for change, and an ideology of high science coupled to high medicine, creating the fine structure for channeling and directing that motive force.

Are we in an analogous position in the late 1980s? Are both of these elements now once again in place? It is probably beyond the scope of my argument to attempt conclusions on the changing climate of medical science. Whether some new set of 'basic sciences,' including bioengineering, decision science, molecular immunology, and (let us say) medical humanities, will become the new intellectual tools to justify new professional arrangements is a question perhaps best left to my colleagues in the policy of health education.

Is American medicine moving into another, quasi-revolutionary period in its ideas and institutions? It is too early to say with any conviction. But it is clear that the motive force for change, the sense of too many doctors with too few hands on the tiller, is very much in evidence. The signs are everywhere, in the pages of almost any general medical journal and in the books and conferences that dot the medical landscape. Some commentators' voices seem shrill and filled with panic, but give us the raw sense of bare-knuckled urgency that is sometimes lacking from the pages of conference volumes. A physician writes, for example, to the *New England Journal* that "slowly but surely the medical profession is allowing itself to be controlled by nonphysicians, who are fast becoming the final authority on medical matters in this country. Lest anyone doubt that this is taking place, the evidence is all around us" [27].

Other commentators provide a more reasoned and moderate response, but prescribe remedies unsurprising to those who have looked at their historical antecedents. One of the best known and most thoughtful of this group, Alvin Tarlov, in his 1983 Shattuck Lecture on the future of physician manpower, noted that

> social systems often evolve slowly, sometimes rapidly. Several social forces intersecting at a critical time cause more rapid change. The health-care system and the practice of medicine have entered a period of rapid change. The forces that are propelling this change are formidable. Whether or not the changes should occur is an interesting but not central question. How to adapt and ensure that the result serves the larger social purpose more effectively is the greater challenge ([39], p. 1243).

Tarlov thus suggested the following measures for coping with the perceived crisis. First, "medical effectiveness should be studied." Outcomes of different services, that is, should be looked at longtitudinally. Second, "a national health-manpower policy should be developed." Coherence should be sought, that is, with an attempt to integrate physician manpower with that of the other health professions. Third, "the growth rate of physician supply should be restrained." That is a familiar drumbeat that we needn't belabor here: "no useful social purpose can be achieved by having more physicians than are needed. Demoralization of the profession will be beneficial to no one" [*ibid.*; I purposely change the order of their presentation].

Last, noted Tarlov, "the doctor-patient relationship should be reconsidered." Doctors and patients, he suggested, "need a conceptual structure to guide their interactions." Until one looks carefully at all the intervening linking arguments, this is a perhaps startling conclusion to draw in a discussion of health manpower. But to the extent that it is a valid suggestion – and I believe it most certainly is – it brings us squarely to that other vertex of the triangle of authority: the cultural authority of the physician.

IV. THE POWER TO NAME

One might begin a discussion of the other, cultural leg of the physician's authority by quoting further from the jeremiad written by the physician who found "nonphysicians . . . fast becoming the final authority on medical matters. . . ." On what, exactly, was she basing her dire prediction of the "takeover of the medical profession," other than the fact that "in the Veterans Administration at least half the members of each hospital clinical executive board . . . are nonphysicians"? Can she identify the engine of this dread institutional dryrot? It seems to lie, by implication, in her concluding paragraph.

> But the final degradation, the ultimate indicator of just how much we have relinquished our power to nonphysicians and to the marketplace, is to have one's medically correct diagnosis changed to a more lucrative but less accurate one by a clerk recording DRGs (diagnosis-related groups). But then DRGs also originated with nonphysicians and were foisted on us by them. Why a once proud and independent profession has permitted this tragic takeover to occur is indeed incomprehensible.

Should we agree with this physician observer? Perhaps in part, if for somewhat different reasons. In the meantime, her views on the physician's potential demotion from helmsman to (at best) assistant navigator are germane, since they revolve around that potent historical source of authority, the power to diagnose.

For many years now, historians, sociologists, and philosophers of medicine have made much of the priestly role of the healer and how that role is reinforced by a body of esoteric wisdom. The power of nosology, the authority of the diagnostician, these are the linchpins of that role: the physician's ability to put a name to the unspeakable enemy within and, by doing so, to demystify that enemy whether or not treatment is promising. But the historical reality, underlying the cultural authority of diagnosis, is more complicated than the simple stereotype of mystification and demystification. Another set of historical vignettes may shed light on this notion.

Though it has many linguistic and biologic roots in traditional western society, modern nosology – the system we use today – goes back no more than a century or, at most, a century and a half. Some time in the middle decades of the last century, a world was gained and another lost. From time to time the physician of today can still be heard pining for commonsensical models of health and disease that may forestall patients' disaffection with modern, "high-tech" medical care [13]. If only physicians can unite with patients in an intuitive, "consensual" view of disease – environmental stress, imbalance of hot and cold, something other than deranged cyclic AMP – then the doctor's cultural authority may be preserved: for what are diseases, after all, but cognitive structures, socially mediated and devoid of inherent reality [11]?

All well and good, but time's arrow points but one way. In an often quoted article, Charles Rosenberg some ten years ago gave historical texture and specificity to the physician's authority to name and, having named, to treat disease. At the beginning of the last century, he noted, with little idea of specific diseases and little more than their raw senses to go on, physicians rendered diagnoses in terms of disruptions of the body's general equilibrium systems. Treatment was immediately thereby dictated, immediately obvious to patient and physician alike, and transparently predictable in its consequences. If something in the body was bottled up, make it flow; if it flowed too freely, bottle it up. Diagnosis and treatment according to this regime allowed for a commodious system in which both doctor and patient could be bound up – what Rosenberg called the "nexus of shared belief and assured relationship." Such a community of belief provided a remarkable degree of emotional and biological satisfaction all around. To a surprising degree it therefore persisted well into the mid-nineteenth century era of therapeutic nihilism, and indeed even into the late nineteenth century era of high science [33].

Modernization, which began with the new pathological anatomy of the Paris hospital [1, 21], and continued with the advent of laboratory medicine, thus brought a trade-off to the physician's power to name. In some sense the new scientific medicine denatured traditional nosology by precipitating it around the specificity of disease rather than the patient. First the histopathological lesion and later the pathogenic organism were emblematic of this change.

And yet diagnosis, both before and after modernization, remained for physicians a process of negotiation, an interpenetration of some hidden biological reality and a cultural setting in which it was manifest and thereby given a name. The designated namer remained the physician. In an important article published a dozen years ago, our commentator today developed this point in a discussion of the "disease of masturbation" [10]. He described the manner in which nineteenth century society, with the physician at its medical helm, divined the "etiology" of the masturbatory syndrome in terms of derangements in nerve tonicity and a state of hyper-excitation. Nineteenth century physicians structured their reality in accordance with their contemporary setting. So, too, did their twentieth century heirs. Masturbation as a correlate of debility gave way to the almost opposite notion of its practice as a corrective for frigidity. In each setting, the author declared,

> expectations concerning what should be significant structure the appreciation of reality by medicine. The variations are not due to mere fallacies of scientific medicine, but involve a basic dependence of the logic of scientific discovery and explanation upon prior evaluations of reality ([10], p. 247).

And in each case the physician reinforced his rightful place as societal helmsman by reading the value system back into the milieu in which he was operating. Again unsurprising: any helmsman who steers his craft aground may be relieved of his watch.

Illustrating the point in another way is chlorosis, a disorder that has received a fair amount of attention from historians in the last decade or so. It is a "defunct disease," like a number of others. We do not, after all, see smallpox any more. But unlike smallpox, chlorosis is *conceptually* defunct. It is unknown to the late twentieth century medical student. The cultural context no longer provides a ready category for it. (Indeed, the Wellcome bibliography pigeon-holes it under "iron deficiency anemia"!)

Chlorosis, the "green sickness," peaked in Victorian America and was a commonplace among women of refinement, especially in their early adulthood and adolescence. Victorian women were often bound into tightly corseted get-ups and shied away from dietary sources of iron such as meat and milk, perhaps in part to curb their otherwise base sexual appetites. One result was this "complex interplay of physiology and social and cultural elements in the production of human pathology," manifested by lassitude, headaches, dyspnea, and other very real somatic symptoms. Clothing and fashion, like moral values, are culturally determined, and it doesn't matter whose theory one subscribes to (for various ideas were advanced to account for the anemia and green pallor of corseted ladies) to explain their biological effects [15].

What does matter is the fact that doctors didn't create chlorosis, nor did their therapies account for its demise. But they did interpret it and mediate the attack upon it, just as they interpreted the disease of masturbation. Today they are called on to interpret and mediate society's approach to disorders like the acquired immune deficiency syndrome (AIDS), said to be a "new disease" of 1980s society, or the post-traumatic stress disorder (PTSD), at least partly a product of the culture of post-Viet Nam America.

Before I conclude I'd like to draw your attention to one final, I think critical, example of the power of diagnosis: the historical impact of bacteriology. The ability of the scientist in the bacteriology laboratory to "make the call," beginning almost exactly one hundred years ago, had extraordinary consequences. Those consequences were felt almost immediately both at the internal boundaries, within the corpus of medicine, and at the external boundaries that defined the relations between physicians and the wider community. The process of moving bacteriology, its conceptual as well as its technical apparatus, into mainstream medicine was more complex than one might intuitively suppose.

On the one hand, there can be no doubt that the specificity and the discriminatory power of bacteriologic diagnosis, as phthisis and consumption gave way to tuberculosis and the acid-fast bacillus, clearly elevated the physician's cultural authority. But there was a price to pay. The pathogenic organisms thought to cause any number of diseases, from beri-beri to mental illness, were confidently isolated and their discovery proudly broadcast.

Within medicine, clinicians, all too aware of the fuzziness and complexity of their task of naming, took the bacteriologists to task for reifying biological taxonomies and reading them into the natural history of human disease. Here was the physician, partly buttressed in authority by the new science, but threatened as well, concerned that the clinical investigator and the doctor at the bedside might be locked into classificatory schemes not of their own devising. Potentially lost, it was felt, was a whole realm of functional diagnosis and symptom complexes – syndromic diagnosis – that traditionally undergirded the authority of the clinical researcher and, by extension, the rank and file clinician [22]. The sociologist and physician Stephen Kunitz has recently expressed this historical tension inherent in the ideology of diagnostics:

> Classifications of disease were not classifications of real entities. The measure of their truth was the pragmatic one of the degree to which they improved the precision of prognosis and therapy, or prediction and control. They could always be changed when new information permitted "better" ways of classifying. Such pragmatic measures of truth seem particularly appropriate for practicing rather than learned professions, for the former act in and on the world whereas the latter are more likely to regard the world contemplatively [20].

The key words in this statement are "prediction and control." For it was the use of diagnosis to impose order, to name and predict, and thereby to establish legitimate control, that ultimately furnished the physician with the necessary cachet and authority. From the 1880s to the 1920s physicians debated long and hard over the predominant language of diagnosis. In the end, enervated, they wound down to something like a draw, since clinicians ultimately *needed* terms like "cardiac failure" as well as terms like "salmonellosis." And ultimately that truce was acceptable from the standpoint of the cultural authority of medicine, since the new, post-Flexnerian configuration of medical education and training made sure to define all the contestants into the same general cultural activity known as "medicine."

V. PARTNERS IN TRUTH-TELLING

But now, and finally coming back to an earlier theme, the physician has a partner in exerting the power of naming, and that partner is by turns the economist, the manager, and the policy scientist. Diagnosis Related Groups (DRGs) have become an established part of the medical care delivery landscape and will also insinuate themselves, in inexorable and some think sinister fashion, into the fabric of medical language as well. The process of establishing a diagnosis rests, and I think will at some level continue to rest, with the physician, but there are new incentives and a complex new set of negotiations to consider. Some commentators are willing to go so far as to make the argument that the advent of DRGs poses the threat, unique in recent times, of not only bureaucratizing the physician's work, but actually proletarianizing the physician him– or herself [20].

In the apocalyptic version of this scenario, the physician's grip is wrested from the tiller and chained to the oars. Whether or not one agrees in detail with this vision, it is hard not to foresee some element of diffusion of responsibility as a potential risk, in a setting where values other than "calling it as one sees it" are at work. Consider what could happen. This is, after all, an age of pleonasm, too many words chasing too little meaning. Here there is no rain without "shower activity" and no cancer without a "mitotic process." It is easy to conceive of administrative personnel and computers gussying up a diagnosis to assure that DRG reimbursements match a patient's hospital stay. We can even conceive of the physician's complicity in this act, especially if his interests are aligned more closely with the institution's success than the patient's.

But, granting all this, does any of it mean the healer need abandon the traditional, pivotal role of truth-teller? In the existential moment between doctor and patient, does pleonasm prevail? Can the physician take his science and the patient's beliefs and distill them, as of old, to that essence of understanding that at once confers comprehension on the patient and reinforced authority on the physician? In individual cases the answer is clearly yes. The problem may be in the aggregate. A new epidemiology of categories, each with its own economic priority, and hence a new gradient of expectations, is created. The issue is not whether managers will ram new and unworkable categories down the throats of unwilling physicians. It is rather the more chilling prospect that physicians will embrace new systems as they come along, with diminishing regard for their truth value. In the immortal words of Pogo, perhaps "we have met the enemy, and they is us," for there are new cadres of physician-planners who are more than pleased to join in the new holy causes of rationalization and efficiency.

If nothing else, the historical record amply demonstrates the plasticity and contingency of disease concepts. Prof. Engelhardt has pointed out how the notion of the "disease entity operates as a conceptual form organizing phenomena in a fashion deemed useful for certain goals"[10]. Now put that together with the "crisis in naming" that currently besets medicine. Lest that be mistaken to be a precious philosophical debate, one need only ponder the call to arms published recently by Smits and McMahon at Yale.

In an article focusing on the capriciousness that rises scum-like to the top of a roiling sea of bureaucracy, Smits and McMahon assess the reasons for all this turbulence. They point out a series of inequities that arise as a result of the interactions between three key elements: the Medicare reimbursement scheme, the DRG classification system that sits at its foundation, and the nosological scheme on which the DRGs in turn are based, known as ICD-9-CM. This last is short for the International Classification of Diseases, 9th revision, Clinical Modification, and it is a tool designed to foster the proper indexing of records and to facilitate retrospective review of medical care. But it is a tool deeply flawed both in structure and in function. If one looks at its structure, for example, one finds that post-infarction angina doesn't exist. Nor does chronic pulmonary disease with infectious complications. One discovers that ulcers treated in any way with an endoscope carry a different code, with double the diagnostic weight (and reimbursement) of even the most difficult and complicated ulcer that is treated medically.

Functionally the current system operates in two curious ways that over time may insidiously skew the way we think about disease. First, Smits and McMahon note, the convention codes for some idealized, crystalline underlying diagnosis rather than its actual real-world manifestations. So a diabetic patient admitted with kidney failure would be given a principal diagnosis of "diabetes with renal manifestations" and assigned to a category of "endocrine disease MDC, 10," with no attempt to discriminate that patient from another who may have been admitted for an attempt to fine-tune his or her glycemic control [24].

Then there is the second functional flaw, namely, the mediation of diagnostic dilemmas, that is, the job of resolving practical questions about the system. The task is relegated to staff members of the American Hospital Association with little or no input from physicians [24]. The result of this ostensible defect, together with the other functional and structural problems of the scheme, is to create a new nosographical map – a new navigation chart, so to speak – that bears some similarity to the old one. But some landmarks are lost, others conflated, and a whole new system of projection appears to have been used.

I make no value judgments as to whether this is "better" or "worse" than before, and word is that, in any event, it is fast becoming an unworkable system, its days numbered. But I do think such attempts at devising bureaucratized nosological systems pose interesting questions about authority. Whoever is communicating with patients and making management decisions is, *de facto*, the captain of the ship. Can the captain do the job optimally with navigation maps that seem at best to have been created by distant committees and at worst to have been traced off funhouse mirrors? This is a question that will be addressed in heated discussion in the next few years.

For now, it would appear that physicians are already dealing with the problem by means of a simple expedient, one that might be suspect in another profession such as accountancy: keeping two sets of books. Or, rather, physicians are keeping two sets of navigation charts. One they learned in medical school. The other is hammered out in committee at places with acronyms (HCFA) that are rapidly becoming more familiar than "CBC" or "SMA 12." And that, to conclude, brings us back to the central conundrum. I think it's fair to suggest that the physician will be asked to remain captain of the ship. Legal authority, though clearly and properly diffused to a degree by "captain of the ship" statutes, will rest in its preponderance and in the majority of instances with doctors. History suggests this to be true, despite the challenges at the legal level swirling out of the so-called crisis of malpractice litigation.

But history also suggests the importance – and the fragility – of those twin buttresses of legal authority, namely, professional and cultural power. Both are now in flux. In the past when this happened there have been major mutations in both the institutional and intellectual arrangements of medicine that represented physicians' (and society's) adaptation to new realities. As an optimist, I believe that is what we will see happening in the next few years. As the captain of the ship struggles with a balky tiller, it seems to me that the law of torts, though it clearly causes financial distortions as insurance premiums mount alarmingly, ultimately does no more than tie one of his pinky fingers behind his back. That is probably not a bad thing.

The other challenges are more serious, with demographic changes threatening to tie the captain's whole professional arm behind his back, and organizational changes threatening his cultural arm as well. But when the dust of doctor-bashing settles down, rationality dictates it to be in the interest of no one to sit someone at the tiller while denying him or her the wherewithal to steer. There will, to be sure, be more navigation aids and more advisers. But to the extent that rationality prevails, the physician will be both forced and properly empowered to remain captain of the ship well into the next century. But the question remains, hanging like the man about to be keelhauled:

Whose rationality is it, anyway?

Presbyterian-University of Pennsylvania Medical Center
Philadelphia, Pennsylvania

NOTES

[1] I wish to thank Kristine Billmyer, Joan Lynaugh, and Steven J. Kunitz for their help as I tried to conceptualize this essay, and for their useful remarks on earlier drafts.

[2] This, I contend – borrowing in part from other critics of Paul Starr's argument [e.g., 34] – is where Starr is at his relatively weakest. In Starr's view [36] the linchpin of cultural authority is societal dependency. The physician has merely capitalized on the decline of a social body that once put its stock in self-reliance, but now staggers on, wallowing in hypochondria and forlorn hopes of a techno-medical quick fix. For my own part, I think the physician's cultural authority relies more straightforwardly on science and less on the infantilization of society. I thus place greater emphasis on the "inner content" of medicine, to which I shall return as well.

BIBLIOGRAPHY

1. Ackerknecht, E.: 1967, *Medicine at the Paris Hospital*, Johns Hopkins University Press, Baltimore.
2. Billings, F.: 1903, '[Presidential Address:] Medical Education in the United States', *Journal of the American Medical Association* **40**, 1271–1276.
3. Brouardel, F.: 1899, *L'exercise de la Médecine et la Charlatanisme*, Baillière, Paris.
4. Brown, T.: 1982, 'An Historical View of Health Care Teams', in G. Agich (ed.), *Responsibility in Health Care*, Reidel, Dordrecht, pp. 3–21.
5. Bullough, V.: 1966, *The Development of Medicine as a Profession*, Karger, Basel.
6. Bylebyl, J.: 1979, 'The School of Padua: Humanistic Education in the Sixteenth Century', in C. Webster (ed.), *Health, Medicine and Mortality in the Sixteenth Century*, Cambridge University Press, Cambridge.
7. Califano, J.: 1986, 'A Revolution Looms in American Health', *New York Times*, March 25, A31.
8. Cipolla, C.: 1976, *Public Health and the Medical Profession in the Renaissance*, Cambridge University Press, Cambridge.
9. Costilo, L.: 1981, 'Competition Policy and the Medical Profession', *New England Journal of Medicine* **304**, 1099–1102.
10. Engelhardt, H.:, 1974, 'The Disease of Masturbation: Values and the Concept of Disease', *Bulletin of the History of Medicine* **48**, 234–248.
11. Fessel, W.: 1983, 'The Nature of Illness and Diagnosis', *American Journal of Medicine* **75**, 555–560.

12. Gelfand, T.: 1980, *Professionalizing Modern Medicine*, Greenwood Press, Westport, Connecticut.
13. Gillick, M.: 1985, 'Common-Sense Models of Health and Disease', *New England Journal of Medicine* **313**, 700–703.
14. Heller, R.: 1978, 'Officiers de Santé: the Second-Class Doctors of Nineteenth-Century France', *Medical History* **22**, 25–43.
15. Hudson, R.: 1977, 'The Biography of Disease: Lessons from Chlorosis', *Bulletin of the History of Medicine* **51**, 448–463.
16. Iglehart, J.: 1986, 'The Future Supply of Doctors', *New England Journal of Medicine* **314**, 860–864.
17. Illich, I.: 1976, *Medical Nemesis*, Pantheon, New York.
18. Ingelfinger, F.: 1976, 'Deprofessionalizing the Profession', *New England Journal of Medicine* **294**, 335.
19. Keel, K.: 1963, *The Evolution of Clinical Methods in Medicine*, Pitman, London.
20. Kunitz, S.: forthcoming, 'Classifications in Medicine', in R. Maulitz and D. Long (eds.), *Grand Rounds: One Hundred Years of Internal Medicine*, University of Pennsylvania Press, Philadelphia.
21. Maulitz, R.: 1987, *Morbid Appearances: the Anatomy of Pathology in the Early Nineteenth Century*, Cambridge University Press, Cambridge.
22. Maulitz, R.: 1979, '"Physician versus Bacteriologist": the Ideology of Science in Clinical Medicine', in M. Vogel and C. Rosenberg (eds.), *The Therapeutic Revolution: Essays in the Social History of American Medicine*, University of Pennsylvania Press, Philadelphia.
23. Mayer, J.: 1985, [Letter], 'A President Argues Against Reducing the Sizes of Medical School Classes', *New England Journal of Medicine* **312**, 1067–1068.
24. McMahon, L. and Smits, H.: 1986, 'Can Medicare Prospective Payment Survive the ICD-9-CM Disease Classification System?', *Annals of Internal Medicine* **104**, 562–566.
25. Moore, F. and Lang, S.: 1981, 'Board-Certified Physicians in the United States: Specialty Distribution and Policy Implications of Trends During the Past Decade', *New England Journal of Medicine* **304**, 1078–1084.
26. Murphy, T.: 1979, 'The French Medical Profession's Perception of its Social Function between 1776 and 1830', *Medical History* **23**, 259–278.
27. Norstrand, I.: 1986, 'Takeover of the Medical Profession by Nonphysicians', *New England Journal of Medicine* **314**, 390.
28. Pelling, M. and Webster, C.: 1979, 'Medical Practitioners', in C. Webster (ed.) *Health, Medicine and Mortality in the Sixteenth Century*, Cambridge University Press, Cambridge.
29. Petersdorf, R.: 1985, 'A Proposal for Financing Graduate Medical Education,' *New England Journal of Medicine* **312**, 1322–1324.
30. Reinhold, R.: 1981, 'The Surgeon General's Pragmatic Boss . . . Efforts to Stimulate Doctor Production "No Longer of High Priority"', *New York Times*, May 19, C3.
31. Relman, A.: 1985, 'Antitrust Law and the Physician Entrepreneur', *New England Journal of Medicine* **313**, 884–885.
32. Rosen, G.: 1944, *The Specialization of Medicine*, Froben, New York.
33. Rosenberg, C.: 1979, 'The Therapeutic Revolution: Medicine, Meaning and Social Change in Nineteenth Century America', in M. Vogel and C. Rosenberg (eds.), *The*

Therapeutic Revolution: Essays in the Social History of American Medicine, University of Pennsylvania Press, Philadelphia.

34. Ruderman, F.:, 1986, 'A Misdiagnosis of American Medicine', *Commentary*, January 1986, 43–49.
35. Schwartz, W. *et al.*: 1980, 'The Changing Geographic Distribution of Board Certified Physicians', *New England Journal of Medicine* **303**, 1032–1038.
36. Starr, P.: 1982, *The Social Transformation of American Medicine*, Basic Books, New York.
37. Stevens, R.: forthcoming, 'Issues in Internal Medicine', in R. Maulitz and D. Long (eds.), *Grand Rounds: One Hundred Years of Internal Medicine*, University of Pennsylvania Press, Philadelphia.
38. Swanson, A.: 1985, 'How We Subsidize "Offshore" Medical Schools', *New England Journal of Medicine* **313**, 886–887.
39. Tarlov, A.: 1983, 'The Increasing Supply of Physicians, the Changing Structure of the Health-Services System, and the Future Practice of Medicine', *New England Journal of Medicine* **308**, 1235–1244.
40. Temkin, O.: 1951, 'The Role of Surgery in the Rise of Modern Medical Thought', *Bulletin of the History of Medicine* **25**, 248–259.
41. U. S. Government: 1975, *Historical Statistics of the United States*, 8, 75–76, Government Printing Office, Washington.
42. Vess, D.: 1975, *Medical Revolution in France, 1789–1796*, Florida State University Press, Gainesville.
43. Weed, L.: 1981, 'Physicians of the Future', *New England Journal of Medicine* **304**, 903–907.

JOAN E. LYNAUGH

NARROW PASSAGEWAYS: NURSES AND PHYSICIANS IN CONFLICT AND CONCERT SINCE 1875

Nursing and medicine are interdependent entities that hold common goals of caring for the sick and curing or preventing illness and share social values of altruism and professional accountability. Still, physicians go about their work of diagnosing and treating disease without giving much thought to nurses; nurses teach patients, clean them, feed them, and support them through illness without paying much attention to physicians. But, as a condition of their practice, nurses and physicians occupy the same space at the bedside of the patient. This shared tenancy may be amicable, tense, or full of outright hostility. Whatever the emotional character of the relationship between nurses and physicians, the relationship itself rests on tradition, beliefs, and laws that stem from ideas of moral responsibility, control, expert knowledge, and correct social order. Any re-examination of the moral and social validity of the role of medicine in contemporary society requires scrutiny of the nurse-physician dyad. It is much easier to understand the issues related to the distribution of work and authority between nursing and medicine when the two disciplines are studied in historical context.

HOSPITAL ORIGINS OF NURSING

The explosive growth of hospitals after the Civil War was a social response to care demands from city dwellers, many of whom were willing to pay for care. These citizens constituted a clientele different from the occupants of the almshouse-hospitals and charity hospitals of the pre-Civil War era. These patients, a mix of working class and a few middle-class people, expected the hospital environment to resemble, at least in some respects, the domestic environment they were used to. The new hospitals, organized and managed by churches, benevolent societies, and ethnic groups, indeed imitated their constituents' homes in design and atmosphere. At first, these new hospitals admitted only adult men and women who required care while sick or undergoing surgical treatment. By the turn of the twentieth century, however, hospitals were admitting children, and a little later they began to open their wards

Nancy M. P. King, Larry R. Churchill, and Alan W. Cross (eds.)
The Physician as Captain of the Ship: A Critical Reappraisal, 23–37.

to obstetrical patients. Over a period of fifty years the idea of the hospital as the right place to be sick and to receive treatment became accepted across a broad spectrum of American society. A hospital became a necessary institution in every community; trained nurses became respectable figures in those institutions.

What was happening was the transplantation of the responsibility for the direct, personal care of the sick from the household to an institution, the hospital. Beyond that, the last quarter of the nineteenth century provided the environment for the delegation of what was once every woman's work, that is, care of the sick, to a special occupational group called trained nurses.

To understand nursing today we have only to search through the images left by about six generations of nurses, most of whom were women. The rapid growth of modern American nursing as a distinct occupation mimics in some senses the spread of the railroads. Both nursing and the railroads developed in a context of rapid industrialization and urbanization and both spread across the country with remarkable speed. There were a few tentative beginnings before the Civil War. These included nursing projects in Philadelphia and Boston; they foretold an explosion of growth which, by 1909, yielded 1,129 new schools for nurses located in some of the 4,700 hospitals ([6], p. 79).

I must spend time discussing hospitals because the invention of community hospitals and the invention of contemporary nursing were simultaneous and interrelated. In other explorations of this topic I have called the hospital initiative of the last half of the nineteenth century the "domestic era" of hospital development [9]. The 1880s and 1890s hospitals often originated in renovated homes; they promised their patients a clean, safe, home-like atmosphere. Usually quite small, they accommodated thirty to one hundred patients and they all faced the same difficult problem. Who would take care of the patients and maintain that necessary atmosphere? In some church-run hospitals there were nuns who could provide this service. In particular, Roman Catholic and Lutheran hospitals held something of a competitive edge, since an easily importable tradition of nursing nuns already existed in Europe. For the other hospitals, however, the lay nurses' training school and nursing supervision of patient care were crucial innovations.

But it would be simplistic to suggest that nursing was invented solely because late 19th century hospitals needed caretakers. Small demonstrations of the social utility of organized groups of trained nurses appeared in the United States as early as 1839. To these was added the moral power of Florence Nightingale's Crimean example, which informed the thinking of American reformers as much as it did their British counterparts. Her book, *Notes On Nursing*, first published in the United States in 1861, influenced many [10]. The depredations of the Civil War focused Americans on the need for caretakers with special knowledge, skill, and discipline.

Nightingale's ideology stressed the primacy of environment and atmosphere in causing disease. She envisioned the hospital as an ordered moral universe that relied on women of high moral character, organizational ability, and a sense of the connectedness between health, behavior, environment, and disease [14]. Trained nurses would "put the patient in the best condition for nature to act upon him" ([10], p. 75]). This set the nurse in the role of manager of human behavior. In this sense the nurse legitimated the idea of clustering sick people together in an institution, since the implicit promise of her presence was to prevent the disorder, chaos, and other unpleasantnesses associated with life inside hospitals [13]. Though Nightingale was by far the best known proponent of nurses as guarantors of the safety and efficacy of hospitals, she was by no means alone in her thinking. In Philadelphia, New Haven, Connecticut, and Boston, physicians were educating women to become nurses who would be "intelligent, benevolent and conscientious" ([11], p. 14). Clearing up the social disorder inherent in illness thus came to rely on the development of a morally superior, domestic institution directed by a strong, practical woman with a sense of vocation.

But what social force drove this sudden enthusiasm for hospitals and for caretakers to substitute for care at home? Historians point to at least four converging factors that seemed to create a need for a new way to care for the sick. First, the location of work was changing from home or farm to a more central location in an office or factory. If anyone in the family became ill or dependent, there may have been no one at home to care for him. Moreover, loss of income resulting from one family wage earner's staying out of work to care for another could inhibit the family prospects. Second, as American families migrated to the cities in the late 19th century, they left behind their extended family and village friends who might help in times of illness. Foreign immigrants, of course, had even less access to assistance if a family member fell ill. In a more speculative vein, I would argue that late 19th century Americans sought more outside intervention when ill than their forefathers. The popular literature of the mid-19th century stressed the therapeutic value of rest away from home, relief from responsibility, and better diet. And, by the 1890s, there seemed to be a growing acceptance of the idea of medical authority as efficacious. Finally, it is quite clear that the advent of antisepsis and anesthesia, which made surgical intervention both safe and endurable, attracted Americans to surgical treatment for a variety of ailments. The hospital of the 1890s was portrayed as a more comfortable and convenient place for surgery than even the most affluent home.

Nurse superintendents, hospital boards, and hospital supporters in each of the domestic-era hospitals tended to be very similar in terms of class, ethnic group, religion, and race. Americans proved willing to experiment with hospitals, but they sought familiar institutions where hospital governors and caretakers would be sensitive to their diet, religion, customs, and language preferences. This match of ethnicity and religion did not apply to hospitals for poor people, the medicalized almshouses, but it was strikingly noticeable in the majority of new community hospitals [1, 9].

Popularity of hospitals also related to ease of transport. Access to the hospital by streetcar and train enabled the patient to get there and allowed his family to visit. Later, the automobile and ambulance further improved the transportation logistics of institutionalizing the sick.

IDEAS OF HEALTH

Even as this new occupation called nursing and these new kinds of institutions were being invented, ideas about health and illness were in flux. Early in the 19th century the body usually was thought of as a system in dynamic interaction with the environment. Health or disease thus seemed to result from interactions between the constitution granted by God to the individual patient and the vagaries of environmental circumstance. Health depended on maintaining one's system in precarious balance between food and liquid taken in and excretions going out. Disease was not visualized as a specifically located entity in the body; instead, illness signaled a general disequilibrium between the body and the environment [7].

This over-all concept began to lose explanatory power as new anatomic, physiologic, and chemical knowledge accumulated. New explanations relying on a more specific concept of disease and complemented by the identification of specific causative agents grew more compelling as the century went on. At the same time, however, in the absence of effective therapy based on these new ideas, caregivers continued to rely on careful instruction and watchful waiting. Indeed, until the twentieth-century advent of sulfa and antibiotics the practice of medicine and the new discipline of nursing both relied heavily on environmental regimens, instruction, and watchful waiting. Physicians, of course, diagnosed disease and prescribed a wide array of drugs. Except for increasingly successful surgical interventions, however, the best that either physician or nurse could do was to follow Nightingale's instruction to "put the patient in the best condition for nature to act upon him" ([10], p. 76).

NURSES AND PHYSICIANS IN THE 19TH CENTURY HOSPITAL

To achieve some sense of how the nurse and physician in the domestic era hospital separated and shared responsibilities I will describe a hypothetical scene. A typical hospital included several large rooms to accommodate eight to twelve patients in a ward, and several smaller rooms for single occupancy. There was a kitchen, a laundry, a drug room, a parlor for admitting patients, receiving guests, and holding board meetings, a room for surgery, a bathroom, sleeping rooms for the nurses, and a chapel if it was a church hospital. Beginning in the 1880s, the caretakers in the hospital included one or two trained nurses (or nuns) and a sufficient number of pupil nurses to care for the patients. The nurse superintendent was appointed by the Board of Trustees and reported to them. She usually ran the hospital as well as the school of nursing with the aid of one or more assistants.

Physicians in the community were free to send their patients to the hospital; some physicians were invited to join the hospital staff. Staff physicians usually provided medical care for the charity patients in the hospital and thus were likely to have more influence there than the non-staff physicians. In particular, surgeons sought to influence the internal operation of the hospital to expedite their surgical work. Importantly, however, neither surgeons nor physicians controlled admissions to these domestic-era hospitals. The closed medical staff and physician control of hospital admissions came later – during the late teens and early twenties of this century.

Community physicians visited the hospital as their patients' care warranted. For the most part, however, physicians practiced in their offices and in the homes of their patients. Nurses and physicians formed working relationships through the hospitals. When the pupil nurses graduated from the hospital training school, physicians might recommend selected graduates to their patients as private duty nurses. Inside the hospital the physician gave instructions to the nurse regarding the patient's diet, medicines, treatments, and activity.

During the domestic era the "hospital as home" metaphor predominated in community hospitals. The nurse superintendent acted as mother in charge and the Board of Trustees acted as father, responsible for finances and over-all decision-making. Physicians, especially those with hospital staff appointments, received respectful attention but were not in a position to control internal events in the hospital.

Hospitals associated with medical schools were exceptions to this pattern, especially those few that were created to serve the teaching and research missions of medical schools. At the Hospital of the University of Pennsylvania and the Johns Hopkins University Hospital, for instance, a quite different relationship existed among trustees, nurses, and physicians. Physicians took a far more active role in planning and decision-making. Similarly, in city hospitals, such as Bellevue and Philadelphia Hospital (later Philadelphia General Hospital), medical faculties used the hospital for teaching young physicians. Those facilities offered a wide array of patients suitable for medical instruction. Here too, physicians involved themselves in hospital governance.

In the majority of hospitals, however, control rested with the trustees; thus, nurses acted as agents of the trustees. Of course, at the same time they accepted instructions from physicians regarding patient care and physicians taught some of the classes in the training school. The domestic-era hospital invented the now familiar triangle of hospital management, nursing, and medicine. The essential balance in the relationships between trustees, nurses, and physicians in the domestic era was to be sharply altered by the escalating interest of physicians in hospitals in the 20th century.

CONFLICTS, SHARED ASSUMPTIONS, AND CHANGE

Significant signs of conflict began to appear at the turn of the century. Two themes pervade this part of the story. First, the hospital became increasingly important to physicians as more and more of their practices relied on hospital-based technology and nursing care. Physicians, as historian Charles Rosenberg puts it, were becoming "hospitalized". The second theme stems from the first: how would nurses allocate their loyalties between trustees and physicians? Whose authority would prevail? And how would nurses act out their responsibilities toward patients in this changing environment?

The "captain of the ship" metaphor appears regularly in medical and hospital literature after 1890, with variations including the "pilot and the wheelman" and "the commanding officer and his lieutenants" [4]. The quasi-military quality of these expressions recalls the pervasiveness and persistence of war-like images in American conceptualizations of health care. The nurses' role in these conceptualizations became that of the disciplined soldier unafraid of contagion, fatigue, or morally dangerous situations. The authority intended to guide her was, of course, the physician. Commentary on the proper relationship between doctor and nurse was an almost inevitable component of speeches given by physicians at nurses' graduation ceremonies at the turn of the century. Historian Susan Reverby calls them "doctor homilies".

As physicians grew more dependent on hospitals, the rhetoric intensified. An illustration of early 20th century dialogue is found in the pages of the 1903 *American Journal of Nursing*. George H. M. Rowe, MD, then Superintendent of the Boston City Hospital, proposed a hospital organization system in which all workers in the hospital, from janitors to the Superintendent of Nurses, would be appointed by a chief executive who would, in turn, be solely responsible to the Hospital Board of Trustees. Such a system would be more efficient, he claimed. Except for the medical staff, which would retain its own autonomy, all other decision-making would be expedited through a single chief. In the next issue, Lavinia Dock, a trained nurse then at the Henry Street Nurses Settlement House in New York City, took Rowe to task for proposing a plan requiring a "subordination [of nurses] which would prevent progress." Dock insisted that the Superintendent of Nurses must retain her direct access to the Board of Trustees. Walking a tightrope between deference and assertiveness, Dock wrote:

> The regard of the well-trained nurse for her own profession and for her professional chiefs, the medical men, is such that she desires for herself a truly dignified position, believing that she will thus best honor her own state, and best deserve the regard of the medical profession ([2], p. 421).

There was both tacit and explicit recognition by nurses of the physician's authority over his patients and over nurses in relation to the medical care of the patients. Nurses tried to draw a distinction between authority over patient care and authority over themselves as persons. But their acceptance of the military metaphor complicated and compromised that distinction. Obedience and loyalty to the nurse in charge and to the physician resonate through all the sets of rules, guide books, ethical statements, and speeches directed at pupil nurses.

Nurses and physicians shared the belief that patients must put complete trust in the physician if they hoped to benefit from care. Nurse leader Charlotte Aiken wrote: "Loyalty to the physician is . . . demanded of every nurse . . . chiefly because the confidence of the patient in his physician is one of the most important elements in the management of his illness . . ." [15]. On the other hand, nurses frequently questioned how far this professional loyalty could go; problems with incompetent physicians and disagreements about what to tell patients abounded. A 1916 formula for dealing with medical errors was offered to nurses by Sara Parsons in her frequently used text *Nursing Problems and Obligations*:

> When she [the nurse] becomes sufficiently experienced to detect a mistake, she will, of course, call his attention to it *by asking if her understanding of the order is correct*. . . . [W]here lives may be at stake each must act as a check upon the other, regardless of an old supposition that doctors could not err and that a nurse had only to obey orders regardless of consequences ([12], p. 58 (emphasis mine)).

Parsons hints that nurse criticism of physicians could be problematic, as she goes on to warn that the nurse "must not allow her work to be underestimated" ([12] p. 58). Fueling the nurses' dilemma was not only their assumption that physician authority and control over patients was vital, but also the assumption that the nurses' own clinical authority depended to some extent on maintaining that of the physician. When nurses could no longer count on delegated power from the lay hospital board members, their dependence on physicians increased.

During the years after 1900, board members in the community hospitals began to relinquish their direct control over hospital affairs to physicians, and nurses found it necessary to accommodate to changing patterns of control. Power relationships in the triangle of hospital management, nursing, and medicine were radically changed by the advent of the closed medical staff in community hospitals. The closed staff system became the norm for community hospitals for two reasons. First, it assured physician staff members of access to hospital beds for their patients, and, second, it assured trustees that their hospital beds would be full, providing, of course, that busy physicians joined the staff. Hospital trustees may or may not have realized how much authority they were giving up when patient admissions became fully controlled by medical staff, but prerogative over admissions gave physicians great power over the economic welfare of the institution and, thus, great influence over all decision-making. In this environment, hospital board members were less able to mediate differences between nurses and physicians effectively.

Thus, I argue that the complicated cooperative and conflicted relationship between nurses and physicians needs to be understood in terms of the environment in which the actors perform. In the earliest hospitals nurses assisted physicians but maintained a domestic form of moral control over their work setting. As physicians found greater need for hospitals to sustain their more technologically complicated medical practice, they also sought more control over the nurses in the hospitals. Nurses acquiesced, to some extent, but tried vigorously to reduce their professional dependence on the hospital, which the physician had so successfully invaded.

NURSING ATTEMPTS TO ACHIEVE PARITY

At least four historical factors help explain nursing's delay in establishing itself as an autonomous entity: its lack of monopolistic control over its work, its reform image as contrasted to medicine's scientific image, its domestic character, which was equated with subservient "women's work", and the vital importance of the training school system to the under-financed but burgeoning hospital industry.

Nurse leaders of the turn of the century recognized the professional peril of their dependence on hospitals as an education base and on physicians as a source of authority. They agitated for university education for nurses, hoping both to exploit the university as a means of attaining professional recognition and credibility and to free their educational programs from the contradictions inherent in their students' roles as apprentices to hospitals no longer under their control. By 1900 the American university was the preferred location for professional training; it ultimately became the pathway through which occupations achieved professional status. Medicine, law, engineering, and even social work established strong links with the university. Nursing, however, failed to achieve this goal until after World War II.

Three generations of nurses sought to extricate their profession from its dependent relationship to hospitals and physicians by seeking endowments for university-based programs. They met with some success; the founding of the Teacher's College nursing program at Columbia University was subsidized by Mrs. Helen Hartley Jenkins, a trustee at Teacher's College. Later, Frances Payne Bolton gave Western Reserve University in Cleveland half a million dollars to start a school of nursing and, in 1923, the Rockefeller Foundation offered grants to Yale and Vanderbilt Universities to establish schools of nursing. But not until massive financial support was generated for nursing education through the G. I. Bill, the National Institute for Mental Health, and the United States Public Health Service after 1945 did nursing education break away from the hospital apprenticeship system into the nation's colleges and universities.

In addition to their constant efforts at educational reform, nurses also found practice opportunities that created distance between them and physicians. Visiting nurse societies, first organized in the late 1880s, began to proliferate in the 1890s. By 1909, a survey counted 1,416 nurses in visiting nursing, revealing that a whole new arena of nursing practice had developed. Nurses were deeply involved in the campaign against tuberculosis and in maternal and child welfare programs. These community nursing care systems offered nurses an important and gratifying role in sanitary and social reform and freed them from constant oversight and direction by physicians. Even here, however, the controversy over authority persisted. As the reform journal *Charities* reported it, the debate focused on "whether the [visiting nurse] is the doctor's nurse, or the patient's nurse; whether the nurse is simply the absolute tool of the physician, doing what he directs and assuming no independent responsibility, or whether, . . . nursing is a distinct profession, with its own legitimate sphere . . ." ([3], p. 580). Fortunately, visiting nurses spent most of their time caring for the poor, which lessened physicians' concern about their activities. Furthermore, visiting nurses practiced in the patients' homes, thereby avoiding physician presence in their immediate practice environment. And, finally, visiting nurse societies were governed by reform-minded Boards of Managers; these affluent women were anxious to actualize ideas of urban improvement and jealously guarded their societies from outside interference.

The vast majority of graduate nurses, however, went into private duty practice after completing training. Though working at a distance from the physicians, they still needed referrals of patients and physician approval of much of their activity. By 1930, nurses practiced in hospitals, visiting nurse societies, or health departments, and in patients' homes doing private duty. Their practice consisted of direct personal care, that is, cleaning, feeding, and helping patients exercise, dressing wounds, and carrying out medically prescribed treatments; they spent time teaching and explaining health matters to patients and their families, they administered drugs, and they tried to sustain patients through illness with whatever interpersonal supports they had to offer.

TWO VIEWS OF CARE

This overview of the American nursing experience depicts a strained but usually cooperative relationship between nurses and physicians, fueled by shared ideas of social responsibility and similar definitions of disease and illness and characterized by interdependence. The relationship is uneasy because of unequal decision-making power between the two professions, social distance, economic inequalities, gender issues, and different priorities in patient care. Conflicts growing out of the gender differences between the professions are, and have been, significant throughout the joint history of medicine and nursing. Conflicts inherent in the now wildly discrepant reward systems are equally significant. These sources of strain and unease are also obvious, at least in the present day. Less well understood, in my view, are the different, though complementary, approaches these health professionals take to their shared patient. Without negating the importance of social inequality and gender differences, it is this latter difference in approach that I want to explore in more detail.

Some time ago a medical colleague and I co-authored a brief essay outlining the language differences between nurses and physicians [8]. For example, what nurses call home visits doctors speak of as house calls. When referring to methods of patient care, nurses discuss the nursing care plan; the doctor orders a medical regimen.

More revealing in terms of the different orientation of the two professions, however, is the language expressing the central question of each profession. Nurses are taught to ask themselves: What are this patient's problems? How is he or she coping with them? What help is needed? What should be left up to the patient? Nursing's phenomena include discomfort, patient worries, mobility, elimination, awareness, and the like. The nursing orientation is inclusive, general, personal, and assistive. Physicians ask themselves: What is this patient's diagnosis? What treatment does he or she need? Medical phenomena include symptoms, disease, therapeutic selection, and therapeutic responses. The medical orientation is selective, disease-focused, objective, and directive.

Over time, nursing has consistently defined its role in society in this general, rather holistic way. Nightingale's definition, "putting the patient in the best condition for nature to work on him," was elaborated in Virginia Henderson's widely accepted 1966 re-statement:

> "The unique function of the nurse is to assist the individual, sick or well, in the performance of those activities contributing to health or its recovery (or to peaceful death) that he would perform unaided if he had the necessary strength, will or knowledge. And to do this in such a way as to help him gain independence as rapidly as possible" [5], p. 15).

Nursing's inclusiveness and its espousal of a supportive, assistive role toward its clients created difficulties and disadvantages in the increasingly reductionist, science-oriented atmosphere of the first half of the 20th century. The biomedical model of disease and cure that swept across the western medical world seemed far more compelling and promising than nursing's holism, environmentalism, and "watchful waiting" approach to illness. Nurses tended to be both inarticulate in explaining their work and touchingly confident that their altruism would eventually be rewarded. The introduction of antibiotics, which reduced the importance of careful nursing in pneumonia, for example, might have presaged the disappearance of nursing as a distinct discipline except for two post-war phenomena.

First, an ever-escalating reliance on complex technology to treat disease and sustain life created an arena of expertise for nursing that nurses successfully exploited during the 1950s and 60s. In 100 years nurses' expertise in ventilation, for example, moved from moving air through hospital wards with draft-free open windows and fireplaces to supporting patients' lives with Bear respirators in critical care units. Second, the lengthening life span of Americans is creating an ever growing population of patients who routinely require health assistance in their daily lives. Some of these are chronically ill persons who survive because of biomedical innovations; some are the aged, who through the attenuations of age come to need nursing expertise to get through their days.

NURSING ENTERS ACADEMIA

When nurses broke through barriers to higher education after World War II, they followed the course laid out by their physician colleagues 60 years before. The crisis of the war, followed by substantial national investment in higher education, opened a window of opportunity for nurses to get into the University system. With federal funding, university schools of nursing flourished and thousands of nurses earned advanced degrees. Equally important, hospital care of patients no longer relied on students; graduate nurses, many with specialty preparation, staffed American hospitals. The combination of more sophisticated education and stable, experienced nurses allowed the development of more and more complex patient care strategies. By the 1970s the pool of nurses with masters and doctoral degrees became large enough to support clinical research, theory development, and testing in addition to maintaining clinical nursing services. A 1984 survey revealed 4,500 nurses holding earned doctoral degrees. They seek and obtain training and research dollars to promulgate and validate their work. They report their work in rapidly multiplying journals and through the public media. And they are making their services more available to the public – in primary care, in industry, in schools, and more recently in nursing homes.

But 75 years of medical dominance over nursing tended to draw a veil over the work of nursing. This invisibility is serious because the result can be that the public's access to nursing care is compromised. In an era where reimbursement for health care is restricted to payment based exclusively on the phenomena of interest to physicians, patients, particularly those who are chronically ill or aged, are deprived of necessary services that nurses can provide. Readjustment of the authority relationship between patient, nurse, and physician comes slowly, step by step. It is profoundly complicated by economic issues, professional territoriality on all sides, and fear of error.

CONCLUSION

It is clear that our system is somewhat askew. The concept of the physician as captain of the ship seems to resonate from another era, though, as history reveals, the authority of the physician-captain, in relation to other caregivers, has always been a subject for debate. Nurses are prepared for and accountable to patients for their practice.

Now they are insisting on recognition for their work. They invest their professional energies in helping patients to cope with health limitations, in teaching, and in direct personal care services. They are not willing, any more than are other professionals, to see their practice curtailed or subsumed in the interest of sustaining the medical profession. Nor are nurses willing to subsidize excessively high incomes for physicians by remaining forever on the economic margins of the health care system.

Turn-of-the-century medical authority was sustained by superior knowledge, control, altruism, and restricted information. These sources of authority are now altered by better general education, delegation of responsibility, rising individualism, and informed consent. Nurses and nursing practice are affected by these social changes as much as physicians. Nurses and physicians remain interdependent in many practice situations and most prefer complementary practice. But it has been a long time since nurses acquiesced to any idea of the physician as in sole control.

The larger question, for both professions, is, how will all health professionals be accountable to their patients? Are the health disciplines capable of responsibly sharing accountability now that the early 20th century myth of physician omnipotence and omnipresence is abandoned?

I believe that nurses and physicians can occupy the same territory; in collaboration the two disciplines provide better, far more effective service to the public than either can alone. Tensions *are* created by proximity; physicians and nurses quarrel occasionally when they jostle each other in the narrow passageways of patient care. Tension, however, is preferable to the distrust and ignorance that stem from silence and distance between the two disciplines. We need to sustain a productive tension. At the same time we need to attain social parity between the two major health disciplines; they must be able to scrutinize and criticize each other. It is far safer for all of our patients that way.

School of Nursing
University of Pennsylvania
Philadelphia, Pennsylvania

BIBLIOGRAPHY

1. Atwater, E.: forthcoming, 'This Worthy Enterprise – The Development of General Hospitals in Upper New York State', in D. Long & J. Golden (eds.), *Hospitals and Communities: A contemporary Institution in Historical Perspective*, Cornell University Press, Ithaca, NY.
2. Dock, L.: 1903 'Hospital Organization', *American Journal of Nursing* **3**, 421.
3. Editorial: 1903, *Charities* **11**, 580.
4. Estes, W. L.: 1891, 'Graduation Address', *Annual Report, St. Luke's Hospital Training School for Nurses*, Allentown, PA.
5. Henderson, V.: 1966, *The Nature of Nursing*, Macmillan Co., New York.
6. *Historical Statistics of the United States, Colonial Times to 1970*: 1975, Washington, DC.
7. Hudson, R.: 1983, *Disease and Its Control*, Greenwood Press, Westport, CT.
8. Lynaugh, J. and Bates, B.: 1973, 'The Two Languages of Nursing and Medicine', *American Journal of Nursing* **73**, 66–69.
9. Lynaugh, J.: forthcoming, 'From Respectable Domesticity to Medical Efficiency: The Changing Kansas City Hospital, 1875–1915', in D. Long and J. Golden (eds.), *Hospitals and Communities: A Contemporary Institution in Historical Perspective*, Cornell University Press, Ithaca, NY.
10. Nightingale, F.: 1860, *Notes on Nursing: What It Is, and What It Is Not*, Harrison & Sons, London (facsimile edition).
11. O'Brien, P.: 'All A Woman's Life Can Bring: The Domestic Roots of Nursing in Philadelphia', *Nursing Research* **36**, 12–17.
12. Parsons, S.: 1985, *Nursing Problems and Obligations*, reprint ed., Garland Publishing, Inc., New York.
13. Reverby, S.: 1987, *Ordered to Care: The Dilemma of American Nursing*, Cambridge Univesity Press, Cambridge.
14. Rosenberg, C.: 1979, 'Florence Nightingale On Contagion: The Hospital As Moral Universe', *Healing and History: Essays for George Rosen*, ed. Charles E. Rosenberg, Science History Publications, New York, pp. 116–136.
15. Winslow, G.: 1984, 'From Loyalty to Advocacy: A New Metaphor for Nursing', *The Hastings Center Report* **14** (3), 32–40.

JUDITH AREEN

LEGAL INTRUSIONS ON PHYSICIAN INDEPENDENCE

It has become fashionable for physicians to blame the reduction in professional autonomy that many are experiencing on lawyers and the legal system [32, 33, 34, 44]. A primary cause of their concern has been significant growth in the cost of medical malpractice insurance premiums, a development that many term a crisis.[1] There are statistics to support the claim of crisis. A recent survey by the American Medical Association (AMA), for example, revealed that the average number of malpractice claims filed annually per 100 physicians increased from 3.3 to 8 in the period from 1978 to 1983 ([32], p. 15). The General Accounting Office (GAO) reports that malpractice insurance costs for physicians doubled between 1983 and 1985, from $1.7 billion to 3.4 billion [43]. There are other statistics, however, that suggest the talk of crisis is exaggerated. The GAO reports that the malpractice insurance premiums paid by the average physician remain "small" in contrast to such other costs as payroll and medical supplies, amounting to about nine percent of the average doctor's total expenses [43].

The conflicting views of whether or not there is a malpractice crisis may reflect the fact that both the incidence of claims and the premium increases have been unevenly distributed. Physicians in certain high-risk specialties, such as neurosurgery, have been particularly hard hit. Some specialists now pay premiums as high as $100,000 annually ([65], p. 10). In 1984 it was reported that 60 percent of all obstetrician-gynecologists had been sued for malpractice, 20 percent of them three or more times ([32], pp. 10–11). Some jurisdictions have experienced higher premium increases as well, particularly Florida, Illinois, Michigan, New York, and the District of Columbia [43].

Economic pressures on physicians stem not only from malpractice claims and insurance premiums, but increasingly from government programs as well. Medicare reimbursement was changed by the Congress in the early 1980's to a prospective payment system under which the amount paid to hospitals is determined by the diagnosis-related group (DRG) into which a particular patient falls [89, 90].[2] The change is feared by some physicians to be the first step toward more fundamental

Nancy M. P. King, Larry R. Churchill, and Alan W. Cross (eds.)
The Physician as Captain of the Ship: A Critical Reappraisal, 39–65.

change in which fee-for-service medicine will be supplanted entirely by state-controlled group health plans.

Federal dollars, as might be expected, brought increased federal oversight of the quality of care. By 1986, federally financed medical review agencies had initiated disciplinary proceedings against more than 740 physicians, charging them with providing poor quality medical treatment to Medicare beneficiaries [36].[3] Another 206 physicians had been charged with providing unnecessary treatment [36].[4] My goal in dwelling on these painful points is not to exonerate my colleagues in the legal profession, for we are certainly not without sin [60]. But I do challenge the perception that the law should be viewed as a primary cause of the problems because the law is for the most part simply an embodiment of choices made by others – including the Congress and state legislatures[5] – who in turn, at least in theory, are acting to protect the interests of the larger society. Blaming "The Law" for the growth in malpractice, or the growing presence of government, is analogous to shooting the messenger who bears bad news. The title of this paper thus should not be read to mean "The Law" is the intruder on physician independence, because the law is merely a stalking horse for others.

Who, then, is the real villain? Paradoxically, it is medicine itself, or at least the successful partnership forged in the twentieth century between science and medicine, that has led to the imposition of significant burdens on the professional autonomy of physicians. The first part of this essay examines three byproducts of the successes of medicine in this century that have fueled legal intrusions on physicians' independence. First, raised public expectations of successful outcomes in medicine have triggered growing demands for accountability for bad outcomes. Second, growing awareness of the limits of medical expertise has prompted a push for increased lay control of treatment decisions. Finally, the escalating costs of health care have generated calls for greater efficiency in the delivery of medical care.

The second part of the essay explores the inconsistencies inherent in the demands for accountability, shared control, and greater efficiency, and analyzes their likely impact on the future of medical practice.

A. THE PRICE OF SUCCESS

There have been such dramatic advances in the ability of physicians to cure or at least to ameliorate illness and injury in this century that it is easy to forget how relatively recently medicine forged its fruitful partnership with science. Lewis Thomas in the title of his autobiography calls medicine "the youngest science" to underscore the point [7]. He recalls, for example, accompanying his physician father on house calls in 1918 when about the only thing a physician could do was diagnose. Little could be done to alter the course of a disease, much less to cure it.[6] Thomas reports that things had not changed very much even by the 1930s when he received his medical education:

> The medicine we were trained to practice was, essentially, Osler's medicine. Our task for the future was to be diagnosis and explanation. Explanation was the real business of medicine. What the ill patient and his family wanted most was to know the name of the illness, and then, if possible, what had caused it, and finally, most important of all, how it was likely to turn out.
>
> The successes possible in diagnosis and prognosis were regarded as the triumph of medical science, and so they were. . . . During the third and fourth years of school we also began to learn something that worried us all, although it was not much talked about. On the wards of the great Boston teaching hospitals – the Peter Bent Brigham, the Massachusetts General, the Boston City Hospital, and Beth Israel – it gradually dawned on us that we didn't know much that was really useful, that we could do nothing to change the course of the great majority of the diseases we were so busy analyzing, that medicine, for all its facade as a learned profession, was in real life a profoundly ignorant occupation ([7], pp. 28–9).

By the second half of the twentieth century, in contrast, medicine's partnership with science and technology had achieved great advances in chemotherapy, operating room hardware, biochemistry, and surgery, to mention only a few areas [68].

This successful partnership not only increased the power and position of physicians and of the medical profession,[7] it also raised expectations of what physicians should be able to accomplish. Today, when the medical treatment provided to particular patients is not successful, it is the many successes of modern medicine that prompt many such patients, or their families, to jump to the conclusion that the physician must have done something wrong.

1. Raised Expectations Trigger Demands for Accountability

In 1982, the Committee on Risk and Decision Making of the National Academy of Sciences reported that despite the fact that by a number of objective measures Americans have never been healthier,[8] most Americans believe life is getting riskier [38].[9] One theory offered to explain the apparent contradiction is that as per-capita income in a society rises, the desire for risk reduction will generally increase. Rising levels of affluence in the United States may be contributing, therefore, to the increased public concern about risk ([38], p. 12). It is as if the traditional "stiff upper lip" had been replaced with a national mood demanding near-total redress for most injuries [59]. Another explanation offered by the Committee is that lofty aspirations reflect faith in the ability of science and technology to provide society with new opportunities to reduce old and emerging risks ([38], p. 12).

Similar factors appear to have influenced public reaction to the practice of medicine. A bad outcome in the past would have been considered unfortunate, but hardly a reason to challenge the competence of the physician. Today, such challenges are becoming commonplace.[10]

Confirmation of the growing public expectation of successful outcomes in medical care, and the corollary demand that physicians be accountable for incompetent performance, can be found in the history of licensure. In the Jacksonian era, as part of the general growth of democratic impulses and its attendant suspicion of privilege, medical licensure (along with legal licensure) came under attack. Illinois abolished licensing in 1826. In Ohio, licensing was introduced in 1811 and repealed in 1833. Licensing laws or penalties for practicing without a license were abolished in Alabama in 1832, Mississippi in 1836, South Carolina, Maryland and Vermont in 1838, and New York in 1844. Several states, including Pennsylvania, never had any licensing ([67], p. 58); [63]).

By the last decades of the nineteenth century, however, the pendulum had swung the other way in many state legislatures, reflecting, no doubt, growing respect for the medical profession, and a concomitant desire to weed out charlatans. The stage was also set for a confrontation in the Supreme Court of the United States over the new licensure laws. In *Dent v. West Virginia* [7], the Court heard a challenge to the West Virginia medical licensing statute brought by a physician claiming that the statute deprived him of the liberty to continue to practice his profession. Justice Field, writing for the Court, upheld the statute in an opinion that emphasized the crucial role of special knowledge and skill in the practice of medicine as justification for public regulation of the profession:

Few professions require more careful preparation by one who seeks to enter it than that of medicine. It has to deal with all those subtle and mysterious influences upon which health and life depend, and requires not only a knowledge of the properties of vegetable and mineral substances, but of the human body in all its complicated parts, and their relation to each other, as well as their influence upon the mind. The physician must be able to detect readily the presence of disease, and prescribe appropriate remedies for its removal. Every one may have occasion to consult him, but comparatively few can judge of the qualifications of learning and skill which he possesses. Reliance must be placed upon the assurance given by his license, issued by an authority competent to judge in that respect, that he possesses the requisite qualifications. Due consideration, therefore, for the protection of society may well induce the State to exclude from practice those who have not such a license, or who are found upon examination not to be fully qualified. . . . No one has a right to practise medicine without having the necessary qualifications of learning and skill; and the statute only requires that whoever assumes, by offering to the community his services as a physician, that he possesses such learning and skill, shall present evidence of it by a certificate of license from a body designated by the State as competent to judge of his qualifications ([7], pp. 122–23).

The opinion also reflects a more general shift in public attitudes that has been summarized by Paul Starr:

[By the last half of the nineteenth century] American politics and culture had undergone a deep change. The American faith in democratic simplicity and common sense yielded to a celebration of science and efficiency. . . . Both the Jacksonians and the Progressives esteemed science, but they understood it in different ways: The Jacksonians saw science as knowledge that could be widely and easily diffused, while the Progressives were reconciled to its complexity and inaccessibility. . . . The assumptions of radicals, reformers and conservatives reflected the more general decline of confidence in the ability of the laymen to deal with their own physical and personal problems. The home medical advisors of the early twentieth century, unlike their predecessors a half century earlier, concentrated mainly on everyday hygiene and first aid. By the Progressive era, to call for popular autonomy in healing was to endanger one's own credibility. The public granted the legitimate complexity of medicine and the need for institutionalized professional authority ([67], p. 140–41).

In the late twentieth century, by contrast, demands for accountability have focused more on individual malpractice claims than on licensure laws. The emergence of the "captain of the ship" doctrine in tort law (a doctrine echoed in the title of this volume) nicely illustrates both the general expansion of medical liability that has occurred in the courts in recent decades and the primary role the law has assigned to physicians in achieving good outcomes.

The phrase "captain of the ship" was first employed in the context of medical malpractice in 1950 in the case of *McConnell v. Williams* [18]. It soon took on a life of its own, however, as was summarized by the Supreme Court of Texas in 1977:

> In naval parlance, the captain of a ship is in total command and is charged with full responsibility for the care and efficiency of the ship and the welfare of all hands. His authority over his own ship and crew is supreme. The captain does not, however, assume personal responsibility for the acts of misconduct or for the criminal deeds committed by the individual men aboard his ship. The court in *McConnell* did not, in fact, impose liability upon the surgeon under its handy phrase which characterized him as the captain of the ship. The court instead ruled that "[i]t is for the jury to determine whether the relationship between defendant and the interns, at the time the child's eyes were injured, was that of master and servant. . . ."
>
> Similes sometimes help to explain a factual situation, but in legal writing, phrases have a way of being canonized and of growing until they can stand and walk independently of the usual general rules. . . . The result in the use of captain of the ship is that a surgeon or physician may be held liable, not as others upon the basis of the general rule of borrowed servant [that is, by showing the physician in fact had authority to control the negligent actions of the health care professional at issue], but as captain of the ship [26].

There has been growing criticism of the doctrine by many legal authorities. Professor Arthur Leff, for example, suggested that it was not so much public faith in the competence of physicians as a search for a "deep pocket" that helped to give birth to the doctrine. He argued, in a posthumously published entry for his dictionary of law, that because the physician is no longer the only available deep pocket, the doctrine should be abandoned:

> Captain-of-the-ship-doctrine. In medical malpractice law, a doctrine which makes surgeons, like ship captains, responsible for the errors of everyone in the operating room, even those who are employed by another, e.g. [the] hospital, and over whose activities they have no actual control and, arguably, no right of control, e.g. anesthesiologists. The doctrine really arose as an attempt to get recompense for the victim: the doctrine of charitable immunity once protected most hospitals, and lack of funds protected most hospital employees; if the surgeon and his insurer couldn't be reached, the savaged patient would be out of luck. With the waning of charitable immunity leading to the procurement by hospitals of comprehensive insurance covering employees, the need for the captain-of-the-ship-doctrine, and its actual application, have similarly decreased [49].

The Surpreme Court of Texas rejected the doctrine outright in 1977 ([26], p. 585). Even in Pennsylvania, the state that gave birth to the doctrine, the courts have now retreated to the view that the doctrine is simply one aspect of the borrowed servant principle that applies generally in the law of agency [28].[11] Nonetheless, the doctrine hangs on in a few jurisdictions (e.g., [30, 31]), underscoring the observation of the Texas Supreme Court that phrases have a way of being canonized. The doctrine's long life, however, probably also reflects judicial awareness of a general public expectation of success in medical care and an associated desire to make physicians accountable for bad outcomes.

2. *The Limits of Medical Expertise and Demands for Shared Authority*

The successes of medicine in the twentieth century have also revealed more clearly to the public the extent to which some choices that need to be made about medical treatment are not medical in nature. The revelation has been particularly dramatic in cases at the end of life that raise the issue of when it is appropriate to terminate or to forego some of the treatment possibilities of modern medicine – chemotherapy, respirator, even artificial nutrition and hydration.

When a patient is competent, a legal consensus has developed in the last decade that the patient has a right to refuse treatment. In *Bartling v. Superior Court* [3], for example, a California appellate court held that a competent adult patient "who was probably incurable but had not been diagnosed as terminal" had the right, over the objection of his physicians and hospital, to have life-support equipment disconnected.[12] In *Bouvia v. Superior Court*, another California court extended the right to refuse treatment to a 28-year-old woman who is quadriplegic and suffering from severe cerebral palsy [4]. In *Estate of Leach v. Shapiro* [8] an Ohio court even held that a cause of action could be brought against a physician for placing and maintaining a patient on life-support systems without informed consent.

The right to refuse treatment has been extended to honoring choices made by formerly competent patients. More than thirty-five states and the District of Columbia have passed living will statutes that enable individuals to put into writing advance directives for health care that are to be followed if they become terminally ill and are unable to direct their own care [66]. In *In re Eichner* [12], also known as the Brother Fox case, the New York Court of Appeals implicitly authorized oral living wills when it permitted testimony as to what Brother Fox, a member of a religious order and a formerly competent patient, had said on the subject of foregoing treatment.

The statutes place a number of restrictions on living wills. Most apply only to patients who are terminally ill. Karen Quinlan, for example, who was diagnosed as being in a persistent vegetative state, was technically not terminally ill, and thus a living will could not have guided her treatment even if she had written one [14]. In some states, moreover, a living will is binding only if the patient has been informed that his or her condition is terminal and then reaffirms the will. In about one-fourth of the states that recognize living wills, they must be reaffirmed periodically to be valid [52].

An increasingly attractive choice for individuals who wish to establish advance directives for health care, therefore, is the use of durable powers of attorney. All fifty states have durable power of attorney statutes that permit a person (termed "principal" in the statutes) to delegate to another the legal authority to act on the principal's behalf [66]. The procedure can be used to delegate authority to a specific person to make decisions, including decisions concerning health care, although there is some debate on this point (compare [66] with [52]).

Decisions to terminate or to forego life-saving treatment are not the only decisions that arise in health care that involve non-medical issues. Decisions such as whether to proceed with elective surgery or which of several treatment alternatives to select have also been increasingly recognized as properly the prerogative of the patient. In the oft-quoted words of Justice (then Judge) Cardozo:

> Every human being of adult years and sound mind has a right to determine what shall be done with his own body; and a surgeon who performs an operation without his patient's consent commits an assault, for which he is liable in damages [25].

In law, the movement to restrict the power of physicians to make choices for their patients has developed primarily through the growth of the doctrine of informed consent. Because that growth has been well chronicled by others [41, 47, 51, 57, 73, 74], I will focus on the most germane recent study of the doctrine.

Professor Jay Katz in his book, *The Silent World of Doctor and Patient* [46], explores the origins of the doctrine in depth. He confirms that there was a link between the ascendancy of science in medicine and the establishment of the concept of informed consent:

> Historically, surrender to silent and blind trust in the physician was to a considerable extent compelled by the state of the art – by medicine's uncertainties that could not be explicated easily. Only during the last 150 years, thanks to the unprecedented advances in medical science, have physicians begun to acquire the intellectual sophistication and experimental tools to distinguish more systematically between knowledge and ignorance, between what they know, do not know, and what remains conjectural. These so recently acquired capacities have permitted physicians to consider for the first time whether to entrust their certainties and uncertainties to their patients ([46], p. xvi).

Katz cautions, however, that although scientific advances have made greater patient participation a possibility, it is not a welcome idea to many physicians. He reports that patient participation is "alien to the ethos of medicine" ([46], p. xvi). He identifies several reasons for this resistance:

> One important reason . . . is physicians' unfamiliarity with and embarrassment over conversing with patients about medical ignorance and uncertainties that can so decisively affect choice of treatment. . . . Revelation of such uncertainties is difficult and disquieting. Learning to live more comfortably with uncertainty, however, has also been impeded by other strongly held, although largely unexamined professional beliefs: that patients are unable to tolerate awareness of the uncertainty, and that faith in professionals and their prescriptions makes a significant contribution to the optimal treatment of disease ([46], p. xvii).

The phrase "informed consent", although foreshadowed by language in earlier judicial opinions, did not appear in case law until 1957, in *Salgo v. Stanford University* [23], a decision by the California Supreme Court. Ironically, the paragraph that gave birth to the doctrine came verbatim (although without attribution) from an amicus curiae brief the American College of Surgeons submitted to the court. Thus, the doctrine so opposed by some physicians was proposed by lawyers in the employ of doctors ([46], p. 60).

Martin Salgo was a 55-year-old man who suffered from leg cramps. Aortography was recommended to see whether he had occlusion of the abdominal aorta. The procedure required injection of a dye, sodium urokon. When Salgo awoke the morning after the procedure, he could not move his legs. He claimed that the permanent paralysis was caused by negligent performance of the aortography. He later added a claim that the physicians negligently failed to warn him of the risks of paralysis inherent in the procedure.

The relevant section of the opinion of the court in *Salgo* begins very forcefully: "A physician violates his duty to his patient and subjects himself to liability if he withholds any facts which are necessary to form the basis of an intelligent consent by the patient to the proposed treatment" ([23], p. 578). But qualifications are soon added: "[I]n discussing the element of risk a certain amount of discretion must be employed consistent with the full disclosure of facts necessary to an informed consent" ([23], p. 578). Unfortunately, no guidance is provided by the court on how to resolve the tension between providing full disclosure on the one hand, and using discretion on the other. That conflict, introduced at the birth of the doctrine, remains central to the current debate on informed consent. Thus, although strong support for patient self-determination can be found in the holding in *Canterbury v. Spence* [5] that disclosure should be judged by what a reasonable patient would want to know, most subsequent court decisions have retreated to the position that physicians should set the standards for disclosure (e.g., [31]). Legislative enactments have followed a similar path ([31], pp. 406–409).

Katz, by contrast, argues there should be "a bona fide attempt by physicians and patients to explain what they wish from one another and what they can do for and with one another, and to clarify, to the extent possible, any misconceptions they may have of each others' wishes and expectations" ([46], p. 125). This is a much more demanding standard for determining whether consent is informed than the law's present narrow preoccupation with disclosure of pertinent risks. Indeed, it may be too subtle a standard for the law to enforce.[13] But it is a standard more in keeping with our emphasis on individual autonomy than is exemplified by the paternalistic tradition, which traces its roots to the Hippocratic corpus that urged the physician to "perform all this calmly and adroitly, concealing most things from the patient while you are attending him" [56]. The time may have come to unseat the traditional paternalistic physician who decides what is best for the patient without consultation or conversation.

The problem of who should decide matters that are beyond medical expertise becomes much harder to resolve when the patient is not competent and did not make known his or her views while competent, or the patient has never been competent because of want of age or mental impairment. Unfortunately, the law on when it is proper to forego or to terminate treatment for incompetent patients is not settled, but varies from state to state. The first court opinions on the subject held that all decisions to withhold or withdraw treatment from incompetent patients should be made by courts, thus further curtailing the traditional authority not only of physicians, but of families as well.

In 1977, for example, the Supreme Judicial Court of Massachusetts issued an opinion in the case of Joseph Saikewicz [27]. He was sixty-seven and had been profoundly retarded all of his life. Although he had two living sisters, neither was willing to be involved in decisions about his care. He had been institutionalized for more than fifty years. He was diagnosed as having acute myeloblastic monocytic leukemia. Evidence before the court indicated there was only a thirty to forty percent chance that the therapy would produce remission, and the probable period of remission was only two to thirteen months. On the other hand, it was reported that most competent patients in his situation chose chemotherapy.

The court held it was legal not to provide chemotherapy to Mr. Saikewicz. It also held that future cases involving incompetents would need judicial resolution.

In 1981, the New York Court of Appeals issued an opinion concerning John Storar [15]. Mr. Storar, age 52, had also been profoundly retarded all of his life and institutionalized for most of if. His mother, a 77-year-old widow, visited almost daily. In 1979, Mr. Storar was diagnosed as having cancer of the bladder. It was treated with radiation, which produced a remission. By 1980, the cancer again produced blood in his urine; this time his condition was diagnosed as terminal. For several weeks, with the permission of his mother, Mr. Storar received blood transfusion. She then asked that they be discontinued. The hospital took the matter to court. The evidence presented showed that Mr. Storar had a very limited lifespan – three to six months at most. He reportedly found the transfusions so disagreeable that he had to be sedated each time.

On appeal, the New York Court of Appeals concluded that because it was impossible to know what Mr. Storar would do if competent, and because the loss of blood could be treated, the only thing to do was to order treatment. Unfortunately, the position, "when in doubt, treat" ignores the very real suffering that the patient may be experiencing.[14] Taken to its logical extreme, the opinion suggests that patients may only be permitted to die from a fatal condition. Any and all other complications that might be treatable in a narrow, technical sense would have to be treated – kidney failure, respiratory failure, etc. – until death occurs from the fatal condition. Worse, the court gave no apparent weight to the views of the family of the patient as to what would be in the patient's best interests.

Recognition of the fact that judicial proceedings can be both cumbersome and costly (in time as well as money) has fueled interest in finding alternative procedures for making treatment decisions for incompetents. One of the most popular alternatives in recent years has been the establishment of ethics committees in hospitals [39, 50]. As Duff has cautioned elsewhere in this volume, however, reliance on ethics committees, like courts, will mean that decisions made will often be less personal because they will be further removed from the knowledge of both physician and family [40]. Nonetheless, the use of ethics committees is likely to spread because their use is thought to reduce the risk that civil or criminal sanctions will be imposed on physicians who terminate or forego treatment [14]. In fact, the degree of risk reduction will depend both on the composition of a particular committee and the procedures it followed as well as on the authority vested in such committees by legislation or judicial decision.

3. The Push for More Efficient Medical Care

A final price medicine has paid for its successes in this century has been growing concern with the efficient delivery of medical care. Once deprivation of basic medical care for economic reasons became socially unacceptable, the government became increasingly involved in paying for medical care [68]. As many predicted, however, federal funding for the poor and the elderly brought increased federal oversight as well [36]. Federal interest in efficiency also grew as the total federal monies spent on health care rose to unprecedented heights. In 1967, Medicare outlays totalled $3.2 billion; in 1984, $50 billion [72]. In the early 1980's, Congress resorted to a prospective payment system in order to halt the spiraling growth in health care expenditures [89, 90].

Cost containment is not a concern of taxpayers only. By 1980, health care expenditures generally had reached $230 billion, up from $69 billion in 1970, a jump from 7.2 to 9.4 percent of the GNP ([67], p. 380). One result has been growing pressure for cost containment in the private sector as well. This new pressure means that physicians will increasingly face a serious conflict of interest – between their traditional allegiance to their patients on the one hand, and a new allegiance to the economic position of the hospital on the other [53]. At a minimum, the latent tensions between hospital governance and administration on one side, and medical staff on the other, are likely to be exacerbated. Ultimately, there may be denials and revocations of hospital staff privileges based on a physician's failure to contain costs. One expert has already predicted that such actions will be upheld by the courts, at least if the hospital has appropriately-drawn medical staff bylaws [45].

B. CHARTING A NEW COURSE

The first part of this essay argued that the successes medicine achieved in this century have imposed costs as well, costs that have limited the professional autonomy and authority of physicians as society has come to expect three things: (1) physician accountability for unsuccessful outcomes when medical treatment is sought; (2) lay authority to decide when there are choices to be made about which treatment to pursue, or whether to be treated at all; and (3) greater efficiency in the delivery of medical services. Unfortunately, the three expectations may be in conflict with one another. People want successful outcomes in medicine, for example, yet at the same time patients who are not trained in medicine want to have more of a say about whether and how they are treated. One way to increase successful outcomes is to perform more diagnostic tests, yet more tests may increase spending on health care. Unfortunately, it is often the physician who is faced with the problem of trying to reconcile such conflicting demands. Still, some encouraging signs are visible.

1. Alternative Mechanisms of Accountability

Raised expectations have led more patients to challenge bad outcomes by taking their physician to court, although many who could sue still do not. Statistics indicate, moreover, that most malpractice cases that go to court are decided in favor of the physician [32]. Although these numbers are undoubtedly skewed by the fact that physicians settle most cases they are likely to lose before they get to court, they suggest that physicians receive some protection from the legal standard employed in malpractice cases – that doctors are to be penalized only if their conduct falls below the standard of the profession, not simply for bad outcomes. The increase in amounts paid in malpractice cases results more from the size of individual awards, it appears, than from an increase in the total number of cases lost ([32], p. 21). The high transaction costs associated with malpractice awards also mean that much of any money awarded goes to lawyers and expert witnesses rather than to injured plaintiffs.

State legislatures are beginning to correct some of the deficiencies in the malpractice system [32]. In 1975, for example, the California legislature enacted the Medical Injury Compensation Reform Act (MICRA) [79]. The legislation was passed in an extraordinary session of the legislature, apparently in response to dramatic and highly publicized increases in the cost of medical malpractice insurance [54]. The Act simultaneously tackled three problems: (1) inadequate oversight of incompetent physicians by the medical profession; (2) excessive costs of malpractice litigation; and (3) excessive malpractice insurance rates.

In order to strengthen professional oversight of incompetent physicians, a State Board of Medical Quality Assurance was established to oversee fourteen local medical quality review committees. A majority of each committee is composed of physicians or surgeons, but each includes lay members as well. The committees are required to investigate complaints of "unprofessional conduct," which is defined to include gross negligence, incompetence, gross immorality, any act involving moral turpitude, dishonesty, or corruption, and any action or conduct that would have warranted denial of a license to practice (the act uses the term "certificate" rather than license). The committees are also to investigate any judgments or settlements for negligence against a physician in excess of a total of $3,000.

The more stringent oversight process has already shown results. In 1983, California ranked eleventh among the states in a ranking based on the number of actions brought per 1000 doctors [75].[15] More serious sanctions (meaning revocation, suspension, or probation) were imposed in California in absolute numbers (117) than in any other state.[16]

Litigation costs were attacked in MICRA by setting a limit on the contingency fees attorneys may charge. The limits are (1) 40 percent of the first $50,000 of any judgment or settlement; (2) 33 1/3 percent of the next $50,000 recovered; (3) 25 percent of the next $100,000; and (4) 10 percent of the amount of any recovery that exceeds $200,000 ([79], sec. 6146(a)).

In *Roa v. Lodi Medical Group* [21], the California Supreme Court upheld the constitutionality of the contingency fee limit. The court conceded the cap might create a potential conflict of interest between attorney and client by reducing the attorney's incentive to pursue a higher award, but pointed out that such conflicts are inherent in all contingent fee arrangements because the difference in the financial position of the lawyer and the client may make for a complete disparity in their willingness to take a risk on a large recovery as against no recovery at all. In response to the contention that the limits set are so low as to make it impossible for any injured person to retain an attorney to represent him or her, the court noted that the limits are not as low as other fee caps (it cited the 25 percent limit on fees for a Federal Tort Claims Act case, among others). To the argument that the limits bear no rational relationship to the objectives of MICRA, the court answered that the limits are a means of deterring attorneys from either instituting frivolous suits or encouraging their clients to hold out for unrealistically high settlements. In addition, the court observed that the fees cap might have been viewed by the legislature as an appropriate means of protecting the already diminished compensation of plaintiffs that results from the provisions of MICRA that limit awards for non-economic damages. The fact that MICRA addressed several facets of the malpractice problem at once thus played a significant role in insulating the Act from constitutional challenge.

The third prong of MICRA tackled the high costs of malpractice insurance by limiting awards for non-economic damages to $250,000 ([79], sec. 333.2). No limits were placed on a plaintiff's right to recover for all of the economic, pecuniary damages (such as medical expenses or lost earnings) resulting from an injury. This cap, in short, was focused on damages awarded for pain and suffering. In *Fein v. Permanente Medical Group* [9], the California Supreme Court upheld the constitutionality of this section of MICRA as well.

Significantly, the United States Supreme Court in 1985 dismissed appeals in both *Roa* [21] and *Fein* [9] for want of a federal question. Such an action by the Court is considered equivalent to an affirmance on the merits [10, 17], although it does not have the same authority as do decisions rendered after argument and with a full opinion [19]. Vindication of the MICRA approach by the Supreme Court may help to slow the judicial tide that in many states had been running against legislative efforts to limit malpractice recoveries ([6]; [60], p. 20).

Mechanisms for weeding out incompetent physicians that are not based on private malpractice mechanisms are also gaining ground. In July, 1986, the board of trustees of the American Medical Association proposed that all hospitals, before admitting a doctor to their staffs, should be required to consult with the AMA to see if the candidate has been disciplined for incompetence in another state [39]. Eleven of the nation's physician-run boards that certify doctors for medical and surgical specialties now require some form of periodic re-examination of competence [69]. The Governor of New York this past summer proposed that all physicians in the state should be re-evaluated every six years [69]. Dr. Otis Bowen, the Secretary of Health and Human

Services, thinks the New York proposal goes a little too far, but nonetheless supports greater efforts to discipline the five to fifteen percent of the nation's 500,000 physicians who fall below acceptable standards for practice [69].

2. New Lay Authority: Reliance on the Families of Incompetent Patients

Informed consent, for better or worse, is now a solid part of the legal landscape. Fortunately, physicians have increasingly come to recognize that there are medical as well as psychological benefits that can accrue from greater patient participation in medical decision-making. But the problem of making decisions for comatose or incompetent patients remains.

In the past several years, a growing number of state courts and legislatures have authorized families to speak for incompetent patients without resort to court [76]. In 1983, in *In re Colyer* [11], the husband of Bertha Colyer, age 69, petitioned the Supreme Court of the State of Washington for discontinuance of life support.[17] Mrs. Colyer had earlier suffered, cardiopulmonary arrest, which led to massive brain damage. She was being kept alive by a respirator. The evidence indicated that she was in a persistent, vegetative state. The Supreme Court of Washington upheld withdrawal of all life-support systems.

The court then took up the question of how to make such decisions for other incompetent patients. The court noted that "the judicial process [is] an unresponsive and cumbersome mechanism for decisions of this nature," and concluded:

> . . . [W]e hold that judicial intervention in every decision to withdraw life sustaining treatment is not required. . . . In cases where physicians agree on the prognosis and a close family member uses his best judgment as a guardian to exercise the rights of the incompetent, intervention by the courts would be little more than a formality ([11], p. 745–46).

The court did qualify its position by noting that the appointment of a guardian "is a judicial process" ([11], p. 746). It also required that a "prognosis board or committee" with no fewer than two physicians with qualifications relevant to the patient's condition in addition to the attending physician should agree that "there is no reasonable medical probability that the patient will return to a sapient state" ([11], pp. 749-750).

In 1984, the Supreme Court of Florida, in *John F. Kennedy Hospital v. Bludworth* [16], gave even greater deference to family choice. Francis B. Landy was terminally ill when admitted to the hospital. Within two days, he was placed on a respirator. He had suffered permanent brain damage, and was unresponsive. The Florida court held that "the right of a patient, who is in an irreversibly comatose and essentially vegetative state, to refuse extraordinary life-sustaining measures, may be exercised either by his or her close family members" or by a court-appointed guardian. Prior court approval need not be sought, the court stated, although "disagreement among the physicians or family members or evidence of wrongful motives or malpractice may require judicial intervention" ([16], pp. 926–27).

In 1984, the Supreme Court of Washington modified the position it had taken in *Colyer*, acknowledging that it was influenced by the decision in *Bludworth*:

> Language used in *Colyer* . . . can be construed to require the appointment of a guardian before any treatment decision can be made concerning an incompetent patient. This, however, is not the intended result.
>
> . . . We do not feel that guardianship proceedings are a necessary predicate to effective decisionmaking in this type of situation. If the incompetent patient's immediate family, after consultation with the treating physician and the prognosis committee all agree with the conclusion that the patient's best interests would be advanced by withdrawal of life sustaining treatment, the family may assert the personal right of the incompetent to refuse life sustaining treatment without seeking prior appointment of a guardian.
>
> In *Colyer* we stated that guardianship hearings would not be overly burdensome, but upon reflection, the approach that best accommodates these most fundamental societal decisions is to allow the surrogate decision maker, the family, to make the decision free of the cumbersomeness and costs of legal guardianship proceedings ([13], p. 1377).

Most recently, an intermediate Arizona court decided *Rasmussen v. Fleming* [20]. Mildred Rasmussen was a 70-year-old patient who had spent the last six years in a nursing home. She was irreversibly comatose as a result of three strokes, a degenerative neuromuscular disease, and/or organic brain syndrome. Her remaining family, a brother and two sisters, together with her physician, sought authority to stop artificial nutrition, which was being provided through a nasal-gastric tube. The court held that a family member has authority without seeking court approval to exercise such a patient's right to refuse treatment, a right the court held to be protected by both the federal constitution and the Arizona constitution. The court also established a priority list of which family member to consult:

(1) The judicially appointed guardian of the person of the patient if such guardian has been appointed. This should not be construed to require such appointment before a treatment decision can be made. (2) The person or persons designated by the patient in writing to make the treatment decision for him. (3) The patient's spouse. (4) An adult child of the patient or, if the patient has more than one adult child, a majority of adult children who are reasonably available for consultation. (5) The parents of the patient. (6) The nearest living relative of the patient . . . [20].

Although there is a trend discernible in these opinions toward empowering family members to make medical treatment decisions for incompetent family members, it is important to note the limitations implicit in the opinions. First, the decisions all authorize the family to speak *on behalf of* the incompetent patient. The rationale for relying on the family is thus only an extension of the policy embodied in both living-will and durable-power-of-attorney statutes. When the patient has not put his or her wishes in writing, the family is a logical substitute decision maker because it is generally in the best position to know what the patient would want done. The family is not, however, authorized to decide on the basis of what the family wants.

Second, all of the cases discussed involved elderly patients. Additional legal standards may apply when the patient is a newborn and not terminally ill [1]. Since 1984, for example, federal law has defined "child neglect" to include the withholding of "medically indicated treatment" from disabled infants with life-threatening conditions [79].

Eleven state legislatures have also now endorsed reliance on the family at the end of life [78, 81, 83–88, 91–93]. Most provide a list of who is to be consulted comparable to the Arizona court decision. Typically, the physician is directed to turn first to the court-appointed guardian if there is one, then to the patient's spouse, then the adult child (or majority of adult children if more than one), then either parent. New Mexico is more demanding than most. Withdrawal of treatment is authorized only when "all family members who can be contacted through reasonable diligence agree in good faith that the patient, if competent, would choose to forego that treatment" [86].

Physicians should generally welcome this new legal trend toward reliance on the family in these cases. First, many physicians were already accustomed to turning to families, despite the fact that it did not provide them complete protection from subsequent civil or criminal action. Second, the new trend means it will not be necessary to have courts or even ethics committees second-guess decisions by families to withhold or to withdraw treatment made on behalf of incompetent patients (and there are likely to be more of these decisions all the time).[18]

3. The Challenge of Cost Containment

There are few encouraging signs in response to the growing push for cost containment. Perhaps the most troublesome indications of what the new push for efficiency in the delivery of medical care may lead to are the growing reports of "dumping" of poor patients, that is, of patients being transferred without regard to their medical need from private to public hospitals because they do not have the insurance to cover the costs of their treatment at private institutions. Congress took some steps in 1985 to combat the problem, but the new legislation will do little to ameliorate the problem without vigorous enforcement. The law provides that any hospital with an emergency department that participates in the Medicare program must ensure that three requirements are met: (1) an appropriate medical screening is provided to every patient who requests it (or has a request made on his behalf) to determine whether an emergency medical condition exists or whether the patient is in active labor; (2) appropriate treatment is provided to stabilize patients who have emergency medical conditions or are in active labor unless the treatment is refused or an appropriate transfer arranged; (3) no patient in an unstable emergency condition or in active labor is transferred unless (a) the patient agrees or a physician determines that the benefits of the transfer outweigh the risks, and (b) the receiving facility has agreed to accept the patient, has space and qualified personnel available for his treatment, and is provided with medical examination and treatment records from the transferring hospital. Any transfer is to be made by proper personnel using equipment that meets health and safety standards. Any hospital that fails to meet these requirements may have its Medicare participation agreement terminated or be fined up to $25,000 for knowing violations [82].

Concern with cost has also been a factor in accelerating the rise of the corporate enterprise in health services. Unfortunately, the competitive market may be an opponent rather than an ally of cost containment. As Eli Ginzberg has cautioned, when capacity increases, advertising and marketing increase, the system expands, costly services are duplicated and the public may be pushed to consume more health care services than it needs [42]. Of thirteen recent studies of whether for-profit hospitals are run more efficiently than non-profit hospitals, most found that they are not. In 1978 and 1980, investor-owned hospitals incurred higher costs per patient per day in every category examined but one: the cost of operating their facility. At the same time, the for-profit hospitals studied also charged significantly more than the non-profit hospitals: an average of 22 per cent more per admission [77].

The growth of for-profit hospitals may also further curtail physician autonomy. The Institute of Medicine has warned that bonuses offered by enterprising hospitals to staff physicians as an incentive for keeping costs down or raising revenues threaten doctors' obligation to act in their patients' best interest [77]. Starr further cautions:

> [t]he great irony is that the opposition of the doctors and hospitals to public control of public programs set in motion entrepreneurial forces that may end up depriving both private doctors and local voluntary hospitals of their traditional autonomy. . . . [C]ompared with individual practice, corporate work will necessarily entail a profound loss of autonomy. Doctors will no longer have as much control over such basic issues as when they retire. There will be more regulation of the pace and routines of work. And the corporation is likely to require some standard of performance, whether measured in revenues generated or patients treated per hour. Humana offers physicians office space at a discount in buildings next to its hospitals and even guarantees first-year incomes of $60,000. It then keeps track of the revenues each doctor generates. . . . Humana's president is frank about what happens if they fail to produce: "I'm damn sure I'm not going to renegotiate their office leases. They can practice elsewhere" ([67], pp. 446–47).

C. CONCLUSIONS

The widespread demands in our society for physician accountability, for increased lay authority, and for efficiency in the delivery of health care show no signs of abating. Each will undoubtedly contribute to the nature of medical practice in the future, and not necessarily in unacceptable ways. The strengthening of state or federal oversight of the quality of medical care provided by physicians, for example, should not worry competent physicians as long as appropriate oversight mechanisms are in place. Stronger self-policing by the profession should help efforts to reduce the amount of malpractice premiums as well as the incidence of claims. Informed consent is now a fixture of the legal system, and is being extended to family members when patients are unable to speak for themselves. This movement to authorize patients and families to make decisions about medical care may actually be a welcome relief to physicians overburdened by the more paternalistic tradition of practice. Finally, even the quest for more efficiency in the delivery of care could reduce resort to unnecessary tests or procedures.

Thus there appears to be reason to expect that these three potentially inconsistent pressures on physicians may be reconcilable, if some care is taken to do so. For example, part of what made MICRA legally as well as politically acceptable was precisely that: its attempt to address more than one concern simultaneously and with due regard for the relationships among them. Moreover, properly understood these pressures reflect appropriate but heretofore incompletely developed views of the nature and scope of the physician's role. The growing adoption of informed consent in end-of-life decisionmaking not only eases decisional pressures on physicians and reduces the likelihood of excess or unnecessary care, but it also proclaims clearly the limits of the physician's decisionmaking role and the scope of the patient's responsibility and rights.

The pressure for cost containment may, however, be different from the others. It certainly poses more risks at present, and thus far there is little reason to suppose it can be adequately circumscribed without the utmost effort and care. In particular, there is a real danger that the pressure for cost containment will be achieved not through efficiency but through reducing the amount or quality of care available to some patients, particularly the poor. This pressure, together with the rise of corporate control of medicine, could radically transform the role of physicians and further reduce their professional autonomy.

A passage in *Moby Dick* describes how command was divided on Dutch whaling ships several centuries ago between the captain and an officer known as the Specksynder. Literally, the word means Fat Cutter. In those days the captain's authority was restricted to navigation and general management of the vessel, while the Specksynder controlled the whale-hunting department. Melville adds that despite the fact that the success of a whaling voyage depends largely on the harpooners, by the nineteenth century the Specksynder had been reduced to an inferior subaltern of the captain.

The physician as captain of the ship is once again in competition with fat-cutters for control. Unfortunately, if present trends continue, it may be the physician who becomes a mere subaltern to the fat-cutters and the next symposium held on these issues will be entitled the Specksynder as Captain of the Ship.

Georgetown University
Washington, DC

NOTES

[1] Claims of a crisis in medical practice were first heard more than a decade ago. Although some now speak of a second crisis, it is probably more accurate to see the trends of the 1980s as a continuation of those of the 1970s [60]. For more detailed discussions of the earlier crisis period, see [60, 61, 70].

[2] Under the new system, the Health Care Financing Administration reimburses hospitals at the pre-set rate for a patient's DRG classification, regardless of the services actually rendered or the actual length of hospitalization. Thus, there is a strong incentive for hospitals to decrease the total cost of services provided to patients, an incentive that may conflict with the obligation of the individual physician to treat patients in a manner that does not fall below the standards of the profession. *See generally* [55].

[3] The actions stem from the peer review process established by the Peer Review Improvement Act of 1982 (Title I, Subtitle C of the Tax Equity and Fiscal Responsibility Act of 1982) [90], which amended Part B of Title XI of the Social Security Act to establish the Utilization and Quality Control Peer Review Organization (PRO) Program. This program replaced the Professional Standards Review Organization (PSRO) Program originally enacted in 1972. PRO's, like their predecessor PSRO's, review health care services paid for by Medicare to determine whether those services are reasonable, medically necessary, furnished in the appropriate setting, and are of a quality that meets professionally recognized standards [50].

[4] Once a disciplinary proceeding has begun, no sanction may be imposed until it has been reviewed by the Inspector General's office of the Department of Health and Human Services. Possible sanctions include fines and temporary or permanent exclusion from the Medicare program. By January 1986, sanctions against some 180 physicians appeared likely [36].

[5] Courts shape the legal landscape as well. In Canterbury v. Spence [5], for example, the standard of disclosure in actions against physicians for failure to obtain informed consent was modified from what other physicians disclose to what a reasonable patient would want disclosed. Most states have not followed the *Canterbury* court.

[6] "I'm quite sure my father always hoped I would want to become a doctor, and that must have been part of the reason for taking me along on his visits. But the general drift of his conversation was intended to make clear to me, early on, the aspect of medicine that troubled him most all through his professional life; there were so many people needing help, and so little that he could do for any of them. It was necessary for him to be available, and to make all these calls at their homes, but I was not to have the idea that he could do anything much to change the course of their illnesses. It was important to my father that I understand this; it was a central feature of the profession, and a doctor should

not only be prepared for it but be even more prepared to be honest with himself about it" ([7], p. 13).

[7] There is some debate about the extent to which other forces produced the growth in prestige and authority of the medical profession in this century. [*See e.g.*, [65] (criticizing Paul Starr's *The Social Transformation of American Medicine* for downplaying the role of science and overemphasizing the decline of democracy and the pursuit of wealth and power by physicians).] Even Starr concedes science was a major force, however:

> As the main emissaries of science, physicians benefited from its rising influence. The continuing growth of diagnostic skills and therapeutic competence was sufficient to sustain confidence in their authority. And with the political organization they achieved after 1900, doctors were able to convert that rising authority into legal privileges, economic power, high incomes, and enhanced social status ([67], p. 142).

[8] Life expectancy at birth increased in the United States from 47 years in 1900 to 74 years in 1979. Mortality reductions in the 1970s were especially impressive. The likelihood that a person age 65 would live at least another decade increased by 14 percentage points from 1900 to 1970 (from 55 to 69 percent) and then by another 5 percentage points by 1979 (to 74 percent) ([38], p. 2).

[9] One Lou Harris poll found that 78 percent of the public surveyed agreed that "people are more subject to risk today than they were twenty years ago;" only 6 percent thought there was less risk ([36], p. 2).

[10] It appears that only a small percentage of those who could bring tort suits do so. One survey of those who reported "a negative medical care experience" found that only eight percent considered seeking legal advice. A study of potentially compensable events in California in 1974 found only one-fourth resulted in tort claims. A 1978 study estimated that the ratio of claims to incidents of malpractice ranged from 0.1 to 0.3 [60].

[11] The borrowed servant principle is used to hold one person (master) who directs or controls a subordinate responsible for the negligence of the subordinate (servant) even though the subordinate is employed by another. Physicians thus have been held liable for the negligence of a nurse under their direct control and supervision even though the nurse is employed by a hospital [27].

[12] *See also* Tune v. Walter Reed Army Medical Hospital [29] (terminally ill patient has right to demand cessation of life support); Satz v. Perlmutter [24] (competent, 73-year-old who is terminally ill with amyotrophic lateral sclerosis [Lou Gehrig's disease] may order respirator removed); Saint Mary's Hospital v. Ramsey [22] (27-year-old Jehovah's Witness who suffers from kidney disease may refuse a life-sustaining blood transfusion).

[13] For a strong argument for adopting a legal standard that is more protective of patient autonomy than the present standard (which protects against inadequate disclosure only when there is bodily harm) *see* [64].

[14] The court qualified its holding as to pain, but not as to suffering, in footnote 7:

> Whether the presence or absence of excessive pain would be determinative with respect to the continuation of a life sustaining measure need not be reached under the facts of this case.

[15] The data were obtained from the Federation of State Medical Boards. Nationwide, some 563 physicians had their licenses revoked or suspended or were put on probation.

The Commission on Malpractice established by the U.S. Department of Health, Education and Welfare in the early 1970's estimated that 3.6 percent of patients who enter hospitals are injured and that 14.5 percent of these injuries were due to negligence. This would yield an estimate of 203,000 people injured in 1983 (when there were 38.8 million hospital admissions).

A 1982 study based on 5,612 surgical admissions to Boston's Peter Bent Brigham Hospital found 36 adverse outcomes due to error during care. This would yield a national estimate of 136,000 injuries.

Extrapolating from 1984 A.M.A. data for doctor-owned insurance companies for claims paid, there were approximately 164,000 injuries in 1983 for which patients were awarded damages either through settlement or adjudication [55].

[16] Florida was second with 71 serious actions. Nine states (Delaware, Vermont, Montana, Idaho, New Hampshire, Rhode Island, West Virginia, Kansas, and Alabama) and the District of Columbia reported no serious actions at all [75].

[17] His affidavit in support of the petition was quoted by the court in its opinion:

> It is very painful for me and Bertha's family to see her in her current condition. We all love her very much and would like for her to be able to live her final days and pass through this life with dignity, rather than being maintained by artificial means ([11], p. 740).

[18] It was reported by the President's Commission for the Study of Ethical Problems in Medicine and Biomedical and Behavioral Research that although a generation or two ago most Americans died at home, now more than 80 percent of us will die in a hospital or long-term care institution surrounded by the technology of our time [58].

BIBLIOGRAPHY

Cases

1. American Hospital Association v. Heckler, 585 F. Supp. 541 (S.D.N.Y. 1984), *aff'd* No. 84–6211 (2d Cir. Dec. 27, 1984), *aff'd sub nom.* Bowen v. American Hospital Association, 106 S. Ct. 2101 (1986).
2. Barber v. Superior Court, 147 Cal. App. 3d 1006, 195 Cal. Rptr. 484 (Ct. App. 1983).
3. Bartling v. Superior Court, 209 Cal. Rptr. 220 (Cal. App. 2d Dist. 1984).
4. Bouvia v. Superior Court, 179 Cal. App. 3d 1127 (Ct. App. 2d Dist. 1986) *pet. for review denied*, June 5, 1986.
5. Canterbury v. Spence, 464 F.2d 772 (D.C.Cir. 1972).
6. Carson v. Maurer, 424 A.2d 825 (N.H. 1980).
7. Dent v. West Virginia, 129 U.S. 114 (1889).
8. Estate of Leach v. Shapiro, 13 Ohio App. 3d 393, N.E.2d 1047 (Ohio App. 1984).
9. Fein v. Permanente Medical Group, 38 Cal. 3d 137, 211 Cal. Rptr. 368, 695 P.2d 665 (1985), *appeal dismissed*, 106 S. Ct. 214.
10. Hicks v. Miranda, 422 U.S. 332 (1975).
11. In re Colyer, 99 Wash. 2d 114, 660 P.2d 738 (en banc) (1983).
12. In re Eichner, 420 N.E. 2d 64 (N.Y. 1981).
13. In re Guardianship of Hamlin, 102 Wash. 2d 810, 689 P.2d 1372 (1984).
14. In re Quinlan, 70 N.J. 10, 355 A.2d 647 (1976), *cert. denied*, 429 U.S. 922.

15. In re Storar, 52 N.Y.2d 363, 438 N.Y.S.2d 266, 420 N.E.2d 64, *cert. denied*, 454 U.S. 858 (1981).
16. John F. Kennedy Hospital v. Bludworth, 452 So.2d 921 (1983).
17. McCarthy v. Philadelphia Civ. Ser. Comm., 424 U.S. 645 (1976).
18. McConnell v. Williams, 361 Pa. 355, 65 A.2d 243 (1950).
19. Metromedia v. San Diego, 453 U.S. 490 (1981).
20. Rasmussen v. Fleming, No. 2 CA-CIV 5622 (Ariz. Ct. App. June 25, 1986).
21. Roa v. Lodi Medical Group, 7 Cal. 3d 920, 211 Cal. Rptr. 77, 695 P.2d 164 (1985), *appeal dismissed*, 106 Sup. Ct. 214.
22. Saint Mary's Hospital v. Ramsey, 465 So. 2d 666 (Fla. Dist. Ct. 1985).
23. Salgo v. Stanford University, 54 Cal. App. 2d 560, 317 P.2d 170 (1957).
24. Satz v. Perlmutter, 362 So. 2d 160, *aff'd* 379 So. 2d 359 (Fla. 1980).
25. Schloendorff v. Society of New York Hospital, 211 N.Y. 125, 105 N.E. 92 (1914).
26. Sparger v. Worley Hospital, 547 S.W.2d 582 (Tex. 1977).
27. Superintendent of Belchertown v. Saikewicz, 373 Mass. 728, 370 N.E.2d 417 (1977).
28. Thomas v. Hutchinson, 442 Pa. 118, 275 A. 2d 23 (1971).
29. Tune v. Walter Reed Army Medical Hospital, 602 F. Supp. 1452 (D.D.C. 1985).
30. Young v. Carpenter, 694 P.2d 861 (Colo. Ct. App. 1985).
31. Woolley v. Henderson, 418 A.2d 1123 (Maine 1980).

Articles, Monographs and Books

32. American Medical Association Special Task Force on Professional Liability and Insurance: 1984, *Professional Liability in the '80's*: *Report 1*, American Medical Association.
33. American Medical Association Special Task Force on Professional Liability and Insurance: 1984, *Professional Liability in the '80's*: *Report 2*, American Medical Association.
34. American Medical Association Special Task Force on Professional Liability and Insurance: 1985, *Action Plan*, American Medical Association.
35. Areen, J., King, P., Goldberg, S., and Capron, A.: 1984, *Law, Science and Medicine*, Foundation Press, Mineola, New York.
36. Brinkley, J.: 1986, 'Disciplinary Cases Rise for Doctors', *New York Times*, January 20, p. 1 at col. 1.
37. Brinkley, J.: 'Medical Association Acts to Tighten Barriers to Incompetence', July 3, p. A10 at col. 1.
38. Committee on Risk and Decision Making, National Research Council: 1982, *Risk and Decision Making*: *Perspectives and Research*, National Academy Press, Washington, D.C.
39. Cranford, R. and Doudera, A. E. (eds.): 1984, *Institutional Ethics Committees and Health Care Decision Making*, Health Administration Press, Ann Arbor, Michigan.
40. Duff, R.: 1988, 'Unshared and Shared Decision Making, Reflections on Helplessness and Healing', in this volume, pp. 191–221.
41. Faden, R. and Beauchamp, T.: 1986, *A History and Theory of Informed Consent*, Oxford University Press, New York.
42. Ginzberg, E.: 1986, 'The Destabilization of Health Care', *New England J. Med.* **315**, 737.

43. Government Accounting Office: 1986, *Medical Malpractice Insurance Costs Increased But Varies Among Physicians and Hospitals*, Washington, DC.
44. Healey, W. V.: 1986, 'The Disaffected Doctor', *Wall Street J.* July 9, at p. 22.
45. Kapp, M.: 1984, 'Legal and Ethical Implications of Health Care Reimbursement by Diagnosis Related Groups', *Law, Medicine & Health Care* **12**, 245–78.
46. Katz, J: 1984, *The Silent World of Doctor and Patient*, The Free Press, New York.
47. Katz, J: 1977, 'Informed Consent – A Fairy Tale? Law's Vision', *U.Pitt. L. Rev.* **39** 137–74.
48. Leff, A: 1985, 'The Leff Dictionary of Law: A Fragment', *Yale L.J.* **94**, 2113 (1985).
49. McCormick, R: 1984, 'Ethics Committees: Promise or Peril?' *Law, Medicine & Health Care*, September, 150–55.
50. Medicare and Medicaid Programs; Utilization and Quality Control Peer Review Organization: Assumption of Medicare Review Functions and Coordination with Medicaid, 50 Fed. Reg. 15312 (1985).
51. Meisel, A: 1979, 'The Exceptions to the Informed Consent Doctrine: Striking A Balance Between Competing Values in Medical Decisionmaking', *Wis. L. Rev.* 413–88.
52. Mishkin, B: 1985, 'Advance Decision-Making for Health Care: Living Wills and Durable Powers of Attorney', in *Taking Charge of the End of Your Life: Proceedings of a Forum on Living Wills and Other Advance Directives*, Wash. D.C.
53. Morreim, E. H.: 1985, 'The MD and the DRG', *Hastings Center Report* **15**, 30–38.
54. Note: 1979, 'California's Medical Injury Compensation Reform Act: An Equal Protection Challenge', *S. Cal. L. Rev.* 52, 829–971.
55. Note: 1985, 'Rethinking Medical Malpractice Law in Light of Medicare Cost-Cutting', *Harv. L. Rev.* **98**, 1004–1022.
56. Pellegrino, E. and Thomasma, D.C.: 1981, *A Philosophical Basis of Medical Practice*, Oxford University Press, New York.
57. Plante, M: 1968, 'An Analysis of Informed Consent', *Fordham L. Rev.* **36**, 639–72.
58. President's Commission for the Study of Ethical Problems in Medicine and Biomedical and Behavioral Research: 1983, *Deciding to Forego Life-Sustaining Treatment*, Government Printing Office, Wash. D.C.
59. Reid, T. R.: 1986, 'Litigation Loosens the Stiff Upper Lip', *Wash. Post*, Feb. 24, p. A1, col. 1.
60. Robinson, G.: 1986, 'The Medical Malpractice Crises of the 1970's: A Retrospective', *Law and Contemp. Prob.* **49**, 5–35.
61. Rottenberg, S. (ed.): 1978, *The Economics of Medical Malpractice*, American Enterprise Institute, Washington D.C.
62. Ruderman, F: 1986, 'A Misdiagnosis of American Medicine', *Commentary*, Jan. 43–49.
63. Shryock, R: 1967, *Medical Licensing in America: 1650–1965*, Johns Hopkins Press, Baltimore.
64. Shultz, M: 1985, 'From Informed Consent to Patient Choice: A New Protected Interest', *Yale Law Journal* **95**, 219–299.
65. Shuchman, M. and Wilkes, M.: 1986, 'Malpractice', *Wash. Post Mag.*, March 2, 6–16.
66. Society for the Right to Die: 1985, *The Physician and the Hopelessly Ill Patient: Legal, Medical and Ethical Guidelines*, New York.
67. Starr, P: 1982, *The Social Transformation of American Medicine*, Basic Books, New York.

68. Stevens, R: 1971, *American Medicine and the Public Interest*, Yale University Press, New Haven.
69. Sullivan, R.: 1986, 'Cuomo's Plan for Testing Doctors is Part of Growing National Effort', *New York Times*, June 9, p. 1 at col. 1.
70. Symposium on Medical Malpractice: Can the Private Sector Find Relief?: 1986, *Law and Contemporary Problems* **49**, 1–348.
71. Thomas, L: 1983, *The Youngest Science: Notes of a Medicine-Watcher*, Viking Press, New York.
72. United States Department of Health and Human Services, Health Care Financing Administration, Health Care Financing Review (Sept. 1983, Dec. 1982).
73. Waltz, J. R.. and Scheuneman, T. N.: 1969, 'Informed Consent to Therapy', *Nw. U. L. Rev.* **64**, 628–650.
74. White, W. D.: 1983, 'Informed Consent: Ambiguity in Theory and Practice', *J. Health, Politics & Law* **8**, 99–119.
75. Wolfe, S., Bergman, H., and Silver, G.: 1985, *Medical Malpractice: The Need for Disciplinary Reform, Not Tort Reform*, Public Citizen, Wash. D.C.
76. Areen, J.: 1987, 'The Legal Status of Consent Obtained from Families of Adult Patients to Withold or Withdraw Treatment,' *J.A.M.A.* **258**, 229 – 235.
77. Light, D.: 1986, 'Corporate Medicine for Profit', *Scientific American* **255**, 38–45.

Statutes
78. Arkansas Death with Dignity Act, Ark. Stat. Ann. Sec. 82–3803 (1977).
79. California Medical Injury Compensation Act (MICRA), Ch. 1, Sec. 1, 1975, Cal. Stats. 2d, Ex. Sess. 3949.
80. Child Abuse Amendments of 1984, Pub. L. 98–457, 98 Stat. 1749.
81. Conn. Gen. Stat. Ann. Sec. 19A–571. (1985).
82. Consolidated Omnibus Budget Reconciliation Act of 1985 (COBRA) Sec. 9121.
83. Fla. Stat. ch. 84–58, Sec. 765–07 (1987).
84. Iowa Right to Decline Life-Sustaining Procedures Act, Iowa Code ch. 144A.7 (1985).
85. Declarations Concerning Life-Sustaining Procedures, La. Rev. Stat. 40: 1299.58.5 (1985).
86. New Mexico Right to Die Act, N.M. Stat. Ann. Sec. 24–7–8.1 (1977).
87. North Carolina Right to Natural Death Act, N.C. Gen. Stat. Sec. 90–322 (1970, amended 1983).
88. Or. Rev. Stat. Sec. 97.083 (1977).
89. Social Security Amendments of 1983, Pub. L. No. 98–21, tit. VI, 97 Stat. 65, 149 (codified in sections of 26 U.S.C. and 42 U.S.C.).
90. Tax Equity and Fiscal Responsibility Act of 1982, Pub. L. No. 97–248, tit. I, 96 Stat. 324 (codified in sections of 26 U.S.C. and 42 U.S.C.).
91. Texas Natural Death Act, Tex. Stat. Ann. art. 4590h, Sec. 4c (1977, amend. 1985).
92. Utah Personal Choice and Living Will Act, Utah Code Ann. Sec. 75–2–1107 (1985).
93. Virginia Natural Death Act, Va. Code Sec. 54–325.8:6 (1983).

H. TRISTRAM ENGELHARDT, JR.

THE AUTHORITY OF THE CAPTAIN: REFLECTIONS ON A NAUTICAL THEME

In many respects, as Areen notes, medicine has never done as well as today. Physicians can intervene with success undreamt of before. With all going so well from a technological and scientific point of view, why all the fuss? Why at times does there seem to be a mutiny brewing? To answer the last question one must recall the various senses in which the physician has been regarded as Captain of the ship. These senses turn on the various ways in which physicians have been seen to be *in* authority or to *be* authorities.[1] I have in mind here a distinction illustrated by one of the differences between sheriffs and lawyers. Sheriffs are *in* authority; they can arrest individuals. Lawyers can be authorities regarding circumstances under which officers of the peace should proceed to arrest individuals. However, lawyers are not as such in authority to arrest individuals. There are numerous ways in which physicians come into authority both within the health care team and with respect to patients. Here it is worth underscoring that even if physicians may be captains of the health care hierarchy or team, it does not follow that they are masters of their patients. One must recall, after all, that the metaphor of physicians as captains of the ship has its strongest originary use in the operating room where the chief surgeon is responsible for his or her assistants as well as the nurses and others aiding in the operation. Even if one held there should be an explicit chain of authority in the operating theater, it would not follow that one held that surgeons ought to have authority over patients beyond that which patients delegated to them. In any event, there are numerous senses of being *in* authority and of being *an* authority. Around each sense there is a special cluster of important issues.

First, recall the patients' rights movement. Concerns with free and informed consent accented the notion that physicians come into authority through the agreement of patients or their families. This contractual view of physician authority contrasts with more paternalistic models where the authority of the physician is derived from the goals of medicine or the special dedication of physicians to the care of patients. The conflict here is between autonomy-oriented and beneficence-

Nancy M. P. King, Larry R. Churchill, and Alan W. Cross (eds.)
The Physician as Captain of the Ship: A Critical Reappraisal, 67–73.

oriented accounts of the moral authority of physicians. But with either interpretation, the moral authority of the physician is not challenged, only the account of its origins. But in either account, the physician's moral authority is seen to come from either implicit or explicit understandings or from certain values supporting the physician's cultural authority to function as either healing priest or scientific technologist, who in naming not only describes, explains, and evaluates reality, but, in addition, creates social reality. This last, so-called gate-keeping function of physicians is like that of sheriffs. Both fashion or shape the character of social relations.

Allied to this first question is a second regarding the delegation of authority within the health care team or the health care hierarchy. Professor Lynaugh's paper addresses this issue. If one holds that the authority of the physician or nurse to treat comes from patients, then one will need to ask how patients see the authority of nurses and other allied health professionals within the health care team or hierarchy. Do they regard their contract with the nurses as with independent contractors? Or do they see themselves giving authority to nurses *through* their physicians in order to maintain an aura of certainty or a special level of coherence in a context of crisis and uncertainty? Do patients want their nurses to be independent professionals, or do they want their nurses to be physician assistants? Do patients want to have consultant physicians under their own authority, or under the authority of the physician of record? These sorts of questions arise in different ways in the operating room and in non-crisis office-based medical care. If authority to treat comes from patients and/or their families, then there will be no answers to discover to these questions. The answers will need to be created by the patients, physicians, nurses, and others involved.

A third set of issues clusters around the ways in which authority is derived from financial transfers, through the exchange of goods for services. When there was true fee for service medicine, physicians brought the money into the system. Physicians could then demand respect from hospitals and thus indirectly from nurses because the physicians were those who kept the system solvent. This point is suggested by Maulitz, and bears being recast here. The increase in the supply of physicians and the development of health maintenance organizations (HMO's) and other structures salarying physicians have undermined this line of authority. So, too, has cost containment. When physicians are not paid either directly by the patient or through a fairly unmonitored retrospective reimbursement system, physicians no longer see themselves beholden only to the patients whom they treat. They also see themselves as responsible to a system of cost containment. This is aggravated even more when utilization review makes it difficult for physicians to admit patients to hospitals and to keep them hospitalized in terms of what appears to the physician to be indicated treatment. Insofar as all-payers systems develop, these changes also undermine the authority of the patient to contract for care. In a system such as Canada, which forbids the sale of private health care insurance, money and goods are devalued in that they can no longer be traded for the provision of health care services through the medium of private health care insurance. In short, the bureaucratization of health care reimbursement has placed a new group of individuals in authority between physicians and hospitals on the one side and patients on the other. As a result, the administrative authority of physicians was challenged by altering the modes and character of reimbursement for services.

Fourth, there have been challenges to restraints by physicians on free trade through licensure laws in the area of health care. The issue of licensure obviously bears on concerns with financial status, prestige, and administrative authority. These questions are joined in different ways by Lynaugh and Areen. To put the issues in perspective, one must note that licensing laws have given physicians the prime legally enforced monopoly on health care services. The license to practice medicine is usually tantamount to the license to practice health care generally. Nurses, optometrists, occupational therapists, and others have been given, at times grudgingly, small areas of their own. Physicians have usually opposed such professionals' practicing independently of control by a physician. Even where such professionals have gained a modest exclave of authority, their area of professional expertise remains a part of the general authority and expertise of physicians. As a result, physicians can always make the claim, with some plausibility, that as psychiatrists, as specialists in physical medicine, as anesthesiologists, or as ophthalmologists, they know what social workers, physical therapists, nurse anesthetists, or optometrists know. One sees this difficulty in the problems that nursing has encountered in defining itself within the academy as an independent discipline, with an integral nursing science, since nursing exists as a single profession in great measure because of the artificial constraints of licensing laws. Often, for example, the experienced intensive care unit (ICU) nurse knows more about treating patients medically, at least in some restricted areas, than the beginning, albeit medically licensed, intern. But the intern has the legal right to prescribe and the nurse does not. As a result, a conflict arises between who is *an* authority, versus who is *in* legal authority. Talk about teamwork can at best blunt some of the tensions and coordinate the actual expertise. The artificial roots of much of these difficulties remain unresolved. In a world without licensure, many nurse practitioners would be practicing what we now call medicine, and the nature of nursing science would become even more obscure.

Fifth, the legal issues plaguing the captain are much more than squabbles among health care professionals regarding legally enforced restraints on trade. The law has intruded ever more into the issues explored under the first two points of this essay – the transfer of authority to treat from patients and/or their families to physicians and the health care team or hierarchy. Areen recognizes this when she refers to the so-called Baby Doe regulations and court decisions, which have inserted state force between the choices of patients and patients' families on the one hand, and their physicians on the other. A recent example is that of the case of Paul Brophy, where the lower court forbade physicians from acting on Brophy's clearly expressed choice not to be treated, with the result that his permanently unconscious body was continued alive over the protests of his family.[2] Legal or state force has thus come to complicate the chains of moral authority. Indeed, the law often wants committees on the bridge. The law has intruded between the choices of the physician and the patients so that physicians can often not be masters of the ship, even if patients and their families demand it.

Sixth, as Areen and Maulitz recognize, the increase in malpractice claims has changed the character of medical practice. I do not know whether or not current levels of litigation are in fact bound to the success of medicine as Areen claims. One should, however, observe that there are simply some cultures that become, for various reasons, highly litigious. One might think, for example, of classical Athens and Iceland. I agree that the problem is in part due to medicine's failure to police itself and in part due to society's unreasonable view that someone is to blame if a patient suffers pain and injury. There has been a failure sufficiently to recognize that much of iatrogenically caused pain and suffering is no one's fault, in the sense of no one's negligence. Rather, all of life is a crap game, medicine is a part of life, and as a result, systems of tort recovery are often expensive ways of insuring against unfortunate mal-events.

But law is not the innocent messenger in all of this, as Areen states. The search for Dr. Deep-pockets, or for that matter Mr. and Mrs. Deep-pockets, is an expression of a mixture of avarice and a failure to develop ways of insuring people against losses without the uncertainties and possibilities of windfall awards integral to the present system. The law as a profession could aid us as a society by attempting to adapt the tort system in ways that would better serve the public good. One might imagine a compulsory review of all suits before trial, with the requirement that plaintiffs who proceeded to trial without a claim that the review panel found to be warranted would need to sustain all court and defense costs, in the event the suit was not successful. One might imagine the bar arguing for abandoning strict product liability in the case of at least vaccines. One might also imagine the bar arguing for limiting recovery in all areas of tort law and providing for structured awards wherever feasible when a stipulated level of insurance had been acquired. Thus one would not only protect against exaggerated awards, but also encourage sufficient protection for those who would need to sue for damages. The law has a responsibility as a learned profession to teach society as well as the members of the legal profession the lineaments of temperance and the counsels of finitude.

In our current circumstances, however, physicians often engage in, or do not engage in, medical procedures because of fears of malpractice. The high level of cesarean sections in the United States probably reflects this fact. Because of the fear of malpractice, physicians at times sail the ship of health care in waters where many patients would not want to enter.

The seventh and next-to-last point raised by the first three essays is more complicated. It bears on the notion of individuals' being authorities, not simply being in authority, or being given authority. Physicians will usually be the medical authorities whether or not they are *in* unambigious legal authority.[3] But as Maulitz suggests, there are also conflicts about who is *the* authority not only between physicians and allied health professionals, but also among physicians themselves. It is often unclear as to whose expertise is truly relevant. One might think, for example, of the classical conflicts between physicians and surgeons regarding whose treatment is indicated and, therefore, who is the relevant expert. These conflicts are often complicated by financial incentives and moral and legal restraints. As a result, it is often a matter to be negotiated, not simply discovered, as to who is *an* authority, as to who is the relevant expert.

This leads to the final point, one suggested by all three papers.

Relationships among physicians, patients, societies, hospitals, allied health professions, and the law are not fixed in stone. They are not outlined in some originary instruction book for the universe. They are created by us. Real men and women decide who is in authority both morally and legally.

Questions about the physician as the Captain of the ship lead us to distinguish various kinds of claims involved in this metaphor. I have tried to bring some order by clustering the issues under eight rubrics. I am sure some could devise further categories. My attempt was not to be exhaustive. On the other hand, my list of eight points could be reduced to more fundamental issues of moral and legal authority, as well as to the powers of expertise and money. My goal was rather to underscore the complexity of the physician's role as Captain or Master of the ship. The extent to which we wish the ship to be one in the Prussian or Texian Navy will depend on values and choices through which we delegate authority and reward expertise. Under the Prussian model, insubordinate nurses may be asked to walk the plank with the approbation of the patients, the physicians, and the hospital. Under the Texian model, the Captain is likely to be challenged to a duel to the death by the nurse anesthestist, while the patients and hospital administrators look on with keen sporting interest. I am not arguing that either model is intrinsically wrong, if those who join enter freely. Rather, I am suggesting that each model has special benefits and banes. Being humans, not gods and goddesses, we can have the benefits of only one model, not all models, at any one time. We had best make our choices with care.

Center for Ethics, Medicine, and Public Issues
Baylon College of Medicine
Houston, Texas

NOTES

[1] For a discussion of various senses of being in authority or being an authority, see [2], p. 105.

[2] In this case the hospital agreed with the court holding. The court injunction forbade the Brophy family from transferring the body of Paul Brophy to a hospital that would be willing to stop all treatment, including artificial hydration and nutrition. "In the event that Patricia E. Brophy, guardian of Paul E. Brophy, causes the ward to be transferred to another medical facility, she is permanently enjoined from authorizing said facility to

either remove or clamp Paul E. Brophy's gastrostomy tube for the purpose of denying said ward hydration and nutrition required to sustain his life" [1].

[3] As Alfred Schutz has shown, the social distribution of knowledge is much more difficult to overcome than the social distribution of money and power. See [3], pp. 324–331. The social constitution of an individual as an expert is a complex process that depends on more than expert knowledge.

BIBLIOGRAPHY

1. Brophy v. New England Sinai Hospital, No. 85E0009–G1 (Mass. Trial Ct., Oct. 21, 1985), *rev'd*, No. N–4152 (Mass. Supreme Judicial Ct, Sept. 11, 1986).
2. Flathman, R.: 1982, 'Power, Authority, and Rights in the Practice of Medicine', in G. Agich (ed.), *Responsibility in Health Care*, Reidel, Dordrecht, Holland, pp. 105–125.
3. Schutz, A. and Luckman, T.: 1973, *The Structures of the Life-World*, trans. R. M. Zaner and H. T. Engelhardt, Jr., Northwestern University Press, Evanston, Illinois.

SECTION II

SHARING THE CAPTAINCY

ERNEST N. KRAYBILL

TEAM MEDICINE IN THE NICU: SHIP OR FLOTILLA OF LIFEBOATS?

The image of a majestic ocean liner plowing the Atlantic, a square-jawed captain firmly gripping the wheel, does not fit the Newborn Intensive Care Unit (NICU). All the NICUs in which I have worked have been small and overcrowded, not at all majestic. The lives that hang in the balance there are also small and fragile. Clearly, the physical symbolism of a ship is not appropriate. What about the organizational metaphor? Is the physician truly captain of the NICU ship? If there is indeed a captain, who are the crew members, who the passengers, and what is their destination? Is the NICU really a ship, or a flotilla of lifeboats?

CHANGING WINDS

Newborn Intensive Care Units have come into existence only in the past quarter century. Alexander Schaffer, who wrote in 1960 what is now regarded as the first textbook of this new field of Pediatrics, asked forgiveness of his readers for coining the words "neonatology" and "neonatologist", which he could not recall having seen in print [6]. Neonatology is a young specialty; its traditions are not so deeply established as those of some other specialties, e.g., surgery, to which the "captain of the ship" metaphor may more accurately apply. Neonatology has been shaped by the times in which it was growing up. The patterns and traditions that do exist have emerged during a period of considerable change in the practice of medicine. This change has been the result of both internal and external forces.

Within the medical profession, an unprecedented amount of biomedical research, funded generously by The National Institutes of Health in the post-World War II years, resulted in an explosion of knowledge that could be applied to the treatment of patients. In the case of newborn infants, an understanding of the pathogenesis of the respiratory distress syndrome of prematurity and an elucidation of the unique metabolic and nutritional needs of infants formed the scientific foundations of the new specialty. Application of this new knowledge was made

Nancy M. P. King, Larry R. Churchill, and Alan W. Cross (eds.)
The Physician as Captain of the Ship: A Critical Reappraisal, 77–88

possible by new technology, particularly the development of instruments for measuring pH and blood gases on micro samples of blood and the refinement of mechanical respirators to meet the needs of tiny infants. These two phenomena, the biomedical research and the technical advances, made possible a lowering of infant mortality in the United States from 26 per 1000 live births in 1960 to 10.6 per 1000 in 1984 [3]. The successful treatment of most infants with respiratory distress syndrome, a disease formerly associated with a high mortality rate, was especially gratifying. As a result of these early successes, neonatology was imbued with a sense of optimism approaching euphoria. The more sobering side of neonatology was, as yet, only dimly perceived.

Outside the profession, changes were occurring as well. The nursing profession, which once was considered, and perhaps even considered itself, to be a "handmaiden to physicians," was seeking a new identity that would place it parallel with, rather than subservient to, medicine. The NICU provided an ideal opportunity to exercise the new consciousness. Other health professions, particularly respiratory therapy, physical therapy, clinical nutrition, and social work, also "discovered" the premature infant and began to develop appropriate roles in the NICU.

The rise of "consumerism" in medicine was felt in neonatology as well as in the entire medical profession. The "doctor knows best" mentality was rejected by parents who wanted to be strongly involved in, if not completely in control of, childbirth, and, in the event of problems, their child's treatment. Other voices were demanding to be heard as well. The concept that a physician's actions must be open to scrutiny, not only by his peers but also by the public, reached its epitome in 1983 when the United States Department of Health and Human Services established toll-free hotlines for anonymous reporting of allegations that handicapped infants were discriminatorily being denied treatment [5].

These and other forces combined to create a milieu that may be unique to NICUs. This background influenced the patterns of authority and responsibility that developed. It seems likely that the personalities of the physicians who invested their professional lives there were also important factors. There is considerable opinion, if not scientific evidence, that pediatricians share personal characteristics that are different from those of surgeons. It is also apparent that the *de facto* organizational structure of an NICU is different from that in an operating room. The patterns of authority, responsibility, and decision-making that I will describe are based on my observations in three NICUs in which I have worked in the past 20 years. They do not necessarily apply to all other NICUs. I shall try to separate clearly my perceptions from a discussion of alternatives. I acknowledge that my perceptions may be biased by my role as Chief of a Division of Neonatal-Perinatal Medicine and my unofficial title "Director of Nurseries".

PASSENGERS AND CREW

To understand the functioning of team medicine in the NICU it is necessary to know something about the passengers, that is, the patients, their illnesses, the available treatments, and the range of outcomes. It is also necessary to have a glimpse of the ship – or the environment of the NICU, the various members of the crew and their functions, the intended destination, and the time frame in which the voyage takes place.

The archtypical NICU patient is a premature infant who was born 12 weeks too early and develops life-threatening respiratory distress syndrome. The parents are shocked, saddened, and sometimes angered by this seemingly unfair blow life has dealt them. Their dreams have been shattered. Instead of the pink, chubby baby they had been expecting, they see a tiny, frail infant, struggling to survive. The respirator, pumps, monitors, tubes, lights, and alarms used to treat the infant are frightening. Unlike the fabled "crisis" of childhood illness a generation ago, when during a brief period of less than 24 hours the child either died or clearly began to recover, the period of uncertainty about a premature infant receiving modern intensive care may go on for months. Not only is the baby at risk of death for a prolonged period; if he survives there is risk of serious permanent handicap. Whether handicap will be present, and if so to what degree, may not be known fully for years. The potential handicaps are not trivial ones. They include blindness, mental retardation, cerebral palsy, seizure disorder, chronic respiratory insufficiency, and growth failure. The goals of newborn intensive care are to permit survival of the infant and to prevent, or at least minimize, permanent disability. It is understandable that the treatment of premature infants often takes place in an atmosphere of deep anxiety. The anxiety affects not only the parents but the staff as well. Occupational stress is high and "burnout" is a common phenomenon in the NICU [2].

Members of a number of medical disciplines work in close proximity in the NICU. Although the responsibilities of each are generally defined, there are areas of overlap. Nurses are central and indispensable – possibly the only truly indispensable group. A nurse usually takes care of no more than two patients at a time. In times of limited staffing, availability of nurses is a controlling factor in admission of patients to the unit.

The nurse carries out a large number of technical and interpersonal tasks. She (98% are female) administers medications and treatments, ensures the baby's comfort and hygiene, observes changes in the baby's condition, records data and provides an important liaison between staff and parents. Most of the important changes in a patient's condition are reported first by a nurse.

A nurse frequently assumes an important role as confidante and counselor to anxious parents. Communications are often on a first-name basis and long-term friendships may develop. Nearly inseparable from that role is the "advocacy" role most nurses fill for their patients. Both of these roles are fostered by a system called "primary nursing," which identifies one nurse who assumes long-term responsibility for each baby. NICU nurses have a sense of identity as members of a highly skilled professional group and have developed local, regional, and national organizations.

Other supporting disciplines, with the exception of social workers, tend to fill roles that are somewhat more completely technical than that of nurses. Respiratory therapists assist in the treatment of the common respiratory disorders. Their role, though still largely technical, has evolved in the NICU, perhaps more than in other hospital units, in a more professional direction. In addition to technical tasks such as setting up, maintaining, and operating respiratory equipment upon order by a physician, well-trained and experienced neonatal respiratory therapists have sufficient understanding of respiratory physiology to make patient assessments and recommend appropriate therapy. Their work, in contrast to that of the NICU nurses, does not encompass all aspects of patient care, but is limited to respiratory illness. Consequently, respiratory therapists are less likely to become acquainted with parents; some patients (those without respiratory illness) are not known to the therapists. Respiratory therapists have not defined an area of practice independent of physicians. By internal as well as external rules, they require a physician's orders for all patient-related activities.

Clinical nutritionists have not yet, as a group, established a consistent role in the NICU. Many NICUs do not have available the services of such specialists. This may be because the nutritional needs of the smallest premature infants have only recently become the subject of extensive scientific inquiry. In NICUs that include a nutritionist on the staff, such individuals may perform educational tasks – teaching nurses and physicians – as well as service – assessing nutritional needs and making recommendations. Although the physicians may translate such recommendations into orders that are carried out by nurses, nutritionists themselves do not write orders, nor do physicians write orders to direct the activities of nutritionists.

Physical therapists, like respiratory therapists, limit their professional activity to one organ system of the premature infant – in this case, the neuromuscular system. They provide expert evaluation and treatment of disorders threatening to limit physical activity. Their role, though not uniformly available in American NICUs, is growing. Because their work is so closely related to child development, physical therapists engage parents in discussions about the baby's progress – an area of vital concern to them. Hence, close personal bonds often develop between therapist and parents.

In contrast to the nurse, respiratory therapist, nutritionist, and physical therapist, the social worker does not deal directly with any of the infant's medical problems but is focused entirely on the parents. Her counseling and advocacy roles are essential to a family's ability to cope with the stress surrounding the prolonged hospitalization of a baby. This function is largely outside, but parallel to, the matrix of day-to-day decision-making that goes on in the NICU. At a deeper level of decision-making, e.g., decisions about the care of a terminally ill infant, the social worker is an essential participant.

Because many NICUs are based in teaching hospitals, interns and residents, often called "house staff", provide much of the medical care. Assignments of house staff to the NICU are on a short-term, rotating basis, usually for a month at a time. During three years of training in pediatrics a resident spends a total of about six months in the NICU. At any given time several residents at differing levels of training work together as a team. There is a hierarchy of responsibility, with more junior physicians being responsible to more senior ones. In many NICUs, there is another layer of medical responsibility assigned to fellows – physicians who have completed general training in pediatrics and are receiving subspecialty training in neonatology. Such physicians function as an extension of the attending staff. Finally, at any given time, and also on a monthly rotation, there is an attending physician to whom all the other staff physicians are responsible. Such an individual has completed two years of subspecialty training beyond general pediatrics training and is certified or eligible for certification in the subspecialty called Neonatal-Perinatal Medicine.

Among physicians in training, there is a wide spectrum, ranging from completely inexperienced physicians who have just recently graduated from medical school to fellows who are nearly prepared to practice independently as neonatologists. The presence in the NICU of inexperienced physicians who are given decision-making responsibility, albeit with close supervision, creates an ironical situation. Inexperienced physicians write orders that direct the activities of highly experienced nurses. This irony reaches its peak when the intern needs to ask the nurse to tell him what orders to write. He is saying, in effect, "tell me what I should tell you to do."

One more kind of crew member needs to be described to understand the operation of the NICU completely. This is the medical or surgical consultant, who may have varying degrees of responsibility for the care of the patient. At one extreme such a consultant may provide a single consultation about a minor problem. At the other extreme, a surgeon may perform a major operation and direct most aspects of postoperative care for an extended period. At times, several consultants may be advising simultaneously about related problems. This has important implications for authority and decision-making.

CHARTING THE COURSE

Somewhere near the center of all this, according to the metaphor, is the attending physician who is in charge of it all – the Captain of the Ship. Is this really the case?

Whatever the reality of authority and responsibility in the NICU, the trappings of traditional physician authority are evident. Doctor's orders, the timehonored institution of medical authority, are central to the operations of the NICU, as in virtually all other units of hospitals. Though usually written by a more junior physician, the orders reflect decisions for which the attending physician is responsible. This is still the method by which prescription is translated into action. A doctor orders a medication or treatment; a nurse or respiratory therapist carries it out. Even in an age of nurse liberation, this institution seems not to have been seriously questioned. To be sure, an individual nurse or therapist may and frequently does challenge a specific order with which she or he disagrees. It would be quite another thing to challenge the system *per se*, and in the hospital setting this has not happened to any significant degree. So deeply entrenched is this system that it is hard even to imagine another one in which nurses wrote orders for physicians, or in which there were no orders.

Yet there are curious inconsistencies and exceptions to the general rule that physicians write orders which are carried out by others. Orders in the NICU are directed, almost exclusively, to nurses and respiratory therapists. Physical therapists somehow enjoy more professional autonomy. While they too must (by rules imposed by the medical profession!) work under a doctor's order, the order is generally non-specific ("Evaluate and treat motor function"). Perhaps this is because physicians understand so poorly the art and science of physical therapy. Social workers enjoy even more professional autonomy, in that they identify the problems to be solved as well as develop a solution. It is not customary for physicians to write orders directing the activities of social workers. This is also true with regard to the work of the clinical nutritionist, whose role is essentially a consultative one.

Even the NICU nurse, ostensibly subject to physician's orders, provides much essential nursing care at her own discretion or based on standards of nursing practice. This includes such things as control of the environment and maintenance of patient hygiene. Other elements of patient care ordered by a physician are subject to modification or cancellation by a nurse, based on her observation and judgment. For example, a formula feeding may be reduced in volume or omitted if an infant shows signs of feeding intolerance.

But what about the essence of decision-making – how does it happen? Whose opinions are heard? Who makes the final decisions, and on what basis? I have already alluded to the phenomenon of the new intern who asks the experienced nurse for advice. Such advice is usually graciously given and gratefully received. But is the nurses' influence limited to the night shift in July and August, when "green" interns may not know how to write feeding orders? It definitely is not. Senior physicians, as well as more junior ones, seek and accept the opinions of experienced nurses on critical issues, including the profound ethical dilemmas which seem to occur with increasing frequency in the NICU. In different and unique ways, the other professionals working in the NICU also contribute to decisions. Respiratory therapists are consulted for their specialized knowledge of respiratory physiology and respiratory equipment. In similar fashion, nutritionists and physical therapists contribute their special knowledge and expertise.

The forum in which suggestions are received and plans developed is another venerable medical institution called "rounds." Though not unique to the teaching hospital, rounds are seen in their fullest flower there, where they take on a teaching function, as well as serving patient care purposes. Indeed, the two functions are so closely intertwined that they cannot be separated. As the cluster of individuals (the team), admittedly physician-dominated, moves from crib to crib, information is received, ideas are debated, and decisions are made. The process is as open or closed as the mind of the individual in charge. In the NICU, the contributions of nurses, respiratory therapists, nutritionists, and social workers are not only welcome, they are actively sought. In this process the whole becomes greater than the sum of the parts. Better patient care results from this team approach, which could be called "team medicine."

"Captain of the ship" seems too restrictive a figure of speech to describe the complexity of relationships that exist in the NICU. In some situations it may fit well – particularly with regard to the treatment of a patient who has a clearly defined illness in need of a specific treatment. In other situations, e.g., when the diagnosis is not clear, when treatment is controversial, or when no effective treatment is available, and when social concerns or moral issues overshadow medical issues, the metaphor may not apply. Captaincy, when it exists, is based on acknowledged competence and a broad perspective, which is necessary to integrate the team members effectively into a common cause. For this reason, captaincy may be transferred as warranted by the situation. For example, a surgeon is often the attending physician of record, and in essence captain, for certain patients who require operations. In other situations, for example, those presenting ethical dilemmas, there may be no acknowledged expert and no logical captain. Yet someone must take responsibility for effecting a course of action. Hospital by-laws invariably vest that responsibility in the attending physician. In the NICU, this is not so clearly a "captain" role; it is perhaps more akin to being a team leader. Professionals working in the NICU are not so much crew as they are team members.

The model of "team medicine" I have described may differ only in degree from the one described by the "Captain of the Ship" metaphor. While perhaps more democratic than the traditional one, it is far from egalitarian. Teams, as well as ships, have captains; the captain of the NICU team is a physician. Physicians still write orders; others, usually nurses, still carry them out. Within the framework of currently accepted medical practice the attending physician has the opportunity to mobilize a wealth of knowledge and experience, which can result in improved patient care. Of course, the ability to seek diverse opinion also implies the right of the physician to reject those opinions in favor of his own. In that sense, the physician, as leader of the NICU team, remains Captain of the Ship.

ALTERNATIVE CAPTAINS, OR NO CAPTAIN

It is possible to conceive of other models. The captain's role could be assigned to an individual in another professional group, for example, a nurse. Or, one could decide not to have a captain, or even a ship, but a flotilla of lifeboats.

I will explore first, and if possible quickly lay to rest, the notion that the NICU does not need a leader – that it should function as a flotilla of lifeboats. Perhaps each professional group could exercise autonomy within its own circumscribed area of practice. Doctors would examine, diagnose, operate, and/or prescribe and administer medicines. Nutritionists would prescribe and administer feedings, while nurses would ensure the comfort and well-being of the patient. Respiratory therapists would both diagnose and treat respiratory ailments. Social workers would counsel the parents. All groups would work with mutual respect but independently. In short, the NICU would not function as a ship or even as a team but as a flotilla of lifeboats.

Such a model would surely result in two or more lifeboat crews working at cross purposes to rescue the same sinking passenger, while failing to notice that some other passengers were not being rescued by anyone. In the unlikely event that such a scheme could be imposed on the professional groups involved, surely it would be unacceptable to parents, society, and the courts, all of which would demand to know: "Who is in charge?"

I shall assume, then, that the NICU team must indeed have a captain, but explore the idea that the captain need not be a physician. In order to assess the validity of any individual or group claim to captaincy, it is necessary first to define the captain's role and establish qualifications. On the ship at sea, the captain has absolute authority and ultimate responsibility. He issues commands that must be obeyed without question. His authority is based on broad knowledge, experience, and demonstrated ability. It is essential that the crew acknowledge those qualities in the captain; otherwise, mutiny would arise.

The nautical model does not fit the NICU. The recent advances in medical knowledge have been so vast that any one individual's knowledge is far from comprehensive. Consequently, no one individual's authority can be absolute. Although a leader is essential, responsibility must be shared. But like a ship's captain, the NICU leader must have broad knowledge, extensive experience, and acknowledged competence. Like the ship's captain, he must effectively blend a wide variety of skills to the common goal – the benefit of the patient. When conflicts arise or when medical caregivers work at cross purposes, the leader must resolve the situation. In doing so, the interests of the patient must have priority over the advancement of any professional group or individual.

The recent efforts by the federal government to direct ethical decision-making in the NICU [5] may be seen by some observers as a bid for the captain's wheel. Ill-advised though those efforts may have been, and though now largely dismantled by the Supreme Court [1], they would not, even had they been successful, constitute captaincy in any sense. Instead, they are analogous to nautical regulations dictating the course by which the captain must guide the ship. Nor does the legitimate involvement of parents in decision-making fulfill the meaning of "captain". They are, rather, an extension of the passengers, for whom the voyage, and indeed the ship, exist. Parents do have both moral and legal authority over the treatment of their own infant, but this does not extend to the broader areas of responsibility implicit in the captain's role. The captain must be concerned for all the passengers, as well as future voyages.

If there is a professional group besides neonatologists that might with some legitimacy lay claim to the wheel of the NICU ship, it could only be nursing. Although I know of no NICU where this has occurred, there is at least one outpatient model of nurse leadership of a multidisciplinary team that includes physicians [4]. No other group besides medicine and nursing is conversant with the broad medical, social, and psychological arenas relevant to the care of the premature infant. Nurses learn to recognize the typical ailments of newborn infants, often before the doctors do. They quickly master many of the technical procedures required for treatment. The nurse's tradition of continuing presence at the bedside may make her a more reliable observer of subtle change than the physician. A nurse may understand, more fully than a physician, the fears and anxieties of the parents, or the pain and discomfort of the patient, or the ethical issues involved. By virtue of her training and personality, the nurse may be more sensitive to the needs and feelings of coworkers, as well as parents. Thus she may be better suited to team leadership than is the physician.

Other considerations would not support the nurse's being designated "captain" of the NICU ship. The most persuasive is that the nurse's training provides neither the breadth nor depth of understanding of physiology, biochemistry, pharmacology, and pathology which may be necessary to integrate the treatment of complex neonatal disease successfully. One might propose that nurses could obtain such training and thus prepare themselves for the captain's role. That would indeed be possible, but then they would have become physicians. More than a few nurses have gone to medical school and become pediatricians – some of them neonatologists and "captains" of NICUs. If one accepts the premise that captaincy is based on competence acknowledged by others, then such training and experience would seem a necessary requisite for a nurse to become captain.

Could a physician provide the necessary professional services and be responsible to a non-physician for direction? There are reasons why such an organizational change is not likely to take place quickly. One that surely must be recognized has to do with human nature. Given medicine's history and traditions, many physicians' self-images would be unlikely to survive such a radical restructuring of professional relationships. A new generation of physicians, perhaps with concepts of authority that differ from traditional ones, may draw on more altruistic motivations to sustain the interest, industry, and scientific inquiry necessary for any field of medicine to advance and thus adapt well to such a change.

Changes in locus of responsibility and authority in the NICU are ultimately determined in large part by the interactions of many forces in American society and only minimally by the medical profession, let alone individual physicians. It was not due to mere chance or whim that public expectations, law, and the values and rules of professional groups, including but not limited to medicine, have created the "Captain of the Ship" model of medical care in the past. Just as surely, those forces have now modified their sanctions to foster what has been described here as the "Captain of the Team" model. Change in medicine, as well as in other institutions, occurs in response to internal and external forces. Although the pace of change has quickened in recent years, it has not been revolutionary. Additional changes will continue to occur, perhaps at an accelerating rate. The direction of change will almost certainly be toward a more democratic, less authoritarian form of medical responsibility. The team is likely to become larger, its functions more complex. The physician may not always be captain. The challenge to health care professionals is to ensure that such changes, when they occur, serve the needs of patients first and the professions second.

University of North Carolina School of Medicine
Chapel Hill, North Carolina

BIBLIOGRAPHY

1. Bowen v. American Hospital Association: 1986, *United States Law Week* **54**, 4579–4594.
2. Clarke T. A., Maniscalco W. M., Taylor-Brown S. T. *et al.*: 1984, 'Job Satisfaction and Stress Among Neonatologists', *Pediatrics* **74**, 52–57.
3. 'Infant and Neonatal Mortality Rates, 1930–84': 1985, *Monthly Vital Statistics Report* **33**, 9.
4. Morris, P.: 1988, 'Who Chartered This Ship?' in this volume, pp.115–124.
5. 'Nondiscrimination On The Basis Of Handicap': 1983, *Federal Register* **48**, 9630–9632.
6. Schaffer AJ: 1960, *Disease of the Newborn*, W. B. Saunders Co., Philadelphia, p. 1.

DAVID BARNARD

"SHIP? WHAT SHIP? I THOUGHT I WAS GOING TO THE DOCTOR!": PATIENT-CENTERED PERSPECTIVES ON THE HEALTH CARE TEAM

People are much greater and much stronger than we imagine, and when unexpected tragedy comes . . . we see them so often grow to a stature that is far beyond anything we imagined. We must remember that people are capable of greatness, of courage, but not in isolation. . . . They need the conditions of a solidly linked human unit in which everyone is prepared to bear the burden of others.

ARCHBISHOP ANTHONY BLOOM

[Natasha's doctors'] usefulness did not consist in making the patient swallow substances for the most part harmful (the harm being scarcely perceptible as they were administered in small doses), but they were useful, necessary, indispensable, because they satisfied a moral need of the patient and of those who loved her, which is why there will always be pseudohealers, wise women, homeopaths, and allopaths. They satisfied that eternal human need for hope of relief, for sympathy, for taking action, which is felt in times of suffering. They satisfied the eternal human need that is seen in its most elementary form in children—the need to have the hurt place rubbed. A child hurts himself and at once runs to the arms of his mother or nurse to have the hurt place kissed or rubbed. He cannot believe that the strongest and wisest of his people have no remedy for his pain. And the hope of relief and the mother's expression of sympathy while she rubs the bump comforts him.

L. TOLSTOY, *War and Peace*

Whatever other virtues it may have, the leading metaphor of this volume is definitely not patient-centered. The image of the physician as the captain of a ship, or of illness and treatment as an ocean voyage requiring the coordination of deck-hands, mates, officers, and nautical machinery, is a bureaucrat's image. It reflects a preoccupation with administrative structure, organizational behavior, and the rational deployment of people and technology. It does not reflect the patient's experience of illness.

This example of the tendency to view the world of health care through the providers' eyes is not surprising. Medicine and its allied disciplines in the humanities and social sciences have long been afflicted with this

Nancy M. P. King, Larry R. Churchill, and Alan W. Cross (eds.)
The Physician as Captain of the Ship: A Critical Reappraisal, 89–111.

disturbance of vision. For most of medicine's modern history, physicians have distrusted patients' views of their own experience [42]. The scientific physician's goal has been to replace the patient's subjective language of distress with data from the laboratory; to translate idiosyncratic or culture-bound expressions of discomfort into the supposedly universal categories of biomedicine. This tendency, which Pfifferling [41] calls "medicocentrism," extends beyond redefining patients' complaints as biophysical derangements. It also includes medicine's selective inattention to the impact of illness and treatment on the patient's daily life, psychological state, or personal values.

Medical sociology and psychology often display medicocentrism in the questions they ask and the presuppositions guiding research. Prime examples are the numerous studies of "compliance" with medical advice that take for granted that the advice was the patient's only rational alternative, was explained thoroughly, and respected the integrity of the patient's belief system. Not surprisingly, with these starting points sociologists find all sorts of characteristics of patients to account for their poor compliance, e.g., negligence, irresponsibility, ignorance, or mistrust. That the "problem" of "non-compliance" is a function of the attitudes, communications behavior, and bureaucratic structures of health care providers themselves could only emerge when the social sciences began to view the system from the patient's perspective [11, 17].

This shift to a more critical stance, and the medicocentrism which it belatedly brought into view, are evoked in a comment by David Mechanic:

> Everywhere in the world the groups least educated, most needy, and suffering from the greatest morbidity are likely to have frames of reference discordant with the assumptions dominant in everyday medical practice. Too frequently, physicians are contemptuous of what they see as superstitious and ignorant attitudes and behavior. Yet medical practice can be conveyed in a fashion consistent with subcultural values and beliefs. Care can be provided so that it is less likely to violate subcultural expectations and norms. The failure to concern itself with the conditions for establishing relationships, inducing cooperation, achieving conformity with suggested regimen, or managing effective follow-up and continuance in care, is a major failure of medicine ([35], p. 21).

Even medical ethics, regarded nowadays as a bastion of the patient's perspective, has a history of medicocentrism. Until very recently, the profession's published codes of ethics have devoted the overwhelming preponderance of their prose to physicians' relationships with each other, and the smooth functioning of hospitals. Obligations to patients run a distant second to these intraprofessional concerns. Jay Katz [25] has argued that contemporary support for informed consent and patient participation in medical decisionmaking is consistently qualified by deference to the physician's judgment about what patients should know and when they should know it. Elsewhere I have argued that unless the medicocentric aspects of everyday clinical behavior are examined for their implicit medicocentric ethic, there is little likelihood that the norms of patient participation and shared decisionmaking will have significant practical effect [1].

My purpose in this essay is to bring to bear a patient-centered perspective on the health care team. I will propose criteria for evaluating health care teams that are based on certain aspects of the patient's experience of illness. These criteria, and the analysis from which they spring, contrast with the bureaucratic, medicocentric criteria that appear most commonly in the professional literature. Whether this contrast also bespeaks a *conflict*, with implications for health care policy, is a separate question to which I will return in my conclusion.

ASSESSING THE HEALTH CARE TEAM

By all accounts the health care team is here to stay. The necessity for team care is part of today's received wisdom. Anyone who questions the *idea* of the team, only to end up reaffirming its importance, is apt to sound like Margaret Fuller, whose pronouncement, "I accept the universe," reportedly provoked Thomas Carlisle to reply, "Gad, she'd better!" Nevertheless, if the importance of the team *per se* is beyond doubt, the choice of criteria for evaluating particular teams remains an open question. Naturally enough, the most popular criteria derive from the arguments that have justified team care in the first place.

These arguments are of three main types: (1) the technical-scientific, (2) the professional-bureaucratic, and (3) the economic. The technical-scientific argument is that since disease is a complex biopsychosocial phenomenon, the required techniques and knowledge bases exceed the capacities of any single practitioner. The professional-bureaucratic argument is simply the institutional extension of the technical-scientific: separate professions have grown up around the various knowledge bases and technologies of modern health care, and their efforts must be brought to bear in a team approach. The economic argument is usually some variation on the theme that the overall cost of caring for complex diseases is reduced through efficiencies deriving from specialization and teamwork, or that teams encourage the utilization of lower-cost preventive and home-based services [19, 37].

These arguments, in turn, suggest technical-scientific, professional-bureaucratic, and economic criteria for assessing health care teams. Evaluation typically attempts to find out whether a particular team has indeed addressed the various components of disease, whether the right mix of professionals is involved and whether their bureaucratic arrangements are smooth or rough, and whether the team is, by some measure, "cost-effective" [3, 50, 57, 59].

A fourth criterion is less frequently mentioned, and far less carefully delineated: how people experience illness. Phenomenologically, people do not experience illness as a congeries of biopsychosocial "aspects." For the patient, as an experiencing subject and as an agent, illness has a phenomenological unity that resists the analytical and professional distinctions underlying the team approach.

An important source of this unity is the social nature of human life. Human beings exist in and for their relationships with others. The stimulus to seek medical aid is, in this sense, always social. It springs from the loss of the ability to perform customary social roles (in one's own eyes or in the eyes of others), from anxiety due to anticipated threats to one's (social) existence, or from the desire to experience in the health care system a form of social contact that is felt to be unavailable in ordinary living [34]. The experiences of falling ill, being ill, seeking help, and recovering from illness are *social events*. They alter the person's sense of connection, reshaping some relationships, creating others, sometimes redefining the conditions under which relationship *per se* can continue as a mode of being-in-the-world [4, 26].

To the bureaucrat, the personnel and institutions that treat disease constitute the "health care delivery system." Increasingly, this system is organized around a proliferating and progressively specialized medical technology. The growth of the allied health professions, and by extension the rationale for the health care team, parallels these technological changes, which require cadres of specialists to perform modern medical work. To the sick person, however, these people and their technologies do not first constitute a "system," but are encountered as changes in the total set of relations that make up his or her being-in-the-world. To be sure, health professionals' technical interventions carry other meanings for the patient that derive from their effects on the patient's physical condition. But they also have an impact on the patient's experience of him- or herself as *being-in-relation* – a point all too easily lost in medicocentric views of health care.

As social events, the ministrations of health professionals help define the sick person's status within the community. From the patient's perspective, they can either reinforce or diminish his or her sense of connection. As Jerome Frank writes,

> The invalid is in conflict with himself and out of harmony with his group. The group is faced with the choice of abandoning him to his fate by completing the process of extrusion, or of making strenuous efforts to heal him, thereby restoring him to useful membership in his community ([15], p. 51).

The social dimension of patients' experience of health care has not been entirely absent from discussion of the health care team. It appears, for example, in these remarks by Avedis Donabedian on assessing quality of care in a health maintenance organization:

> It would be presumptuous of me to remind anyone that, in addition to the quality of technical care, clients are influenced by the nature and stability of the interpersonal relationships with their health care practitioners, by the amenities of the settings in which care is received, and by the ease of access to care when care is thought to be needed. The point that I do want to make is that all of these considerations are potential ingredients in a truly comprehensive and relevant measure of the quality of care. . . .
>
> Satisfaction with these aspects of care casts a glow that colors many other client experiences in receiving care. Similarly, dissatisfaction with the client-practitioner relationship has a chilling effect that no amount of technical expertise seems able to fully overcome. Though clients are initially grateful, especially when some remarkable feat of technical prowess has been accomplished, they later resent bitterly any real or fancied affront they were made to suffer in the process. The more dependent the clients may have been for a while on the practitioner's goodwill, the more they resent it if they believe that advantage has been taken of their helplessness ([12], p. 219).

While these remarks are suggestive, a systematic foundation for assessing the relational aspects of the health care team requires a more thorough analysis of human connectedness, and an effort to relate that analysis to people's experience of illness. I intend to begin by exploring the nature of human connectedness as it appears in the work of John Bowlby on attachment behavior, and in D. W. Winnicott's work on maternal-infant relationships and "the holding environment."

Some of this material may initially seem far afield from health care teams. I will further argue, however, that illness heightens people's need for attachment to others, and insight into the infantile roots of attachment behavior throws important light on the social and moral contours of health care. Deeper understanding of the nature and vicissitudes of human attachment will suggest criteria for health care teams that correspond to people's experience of illness.

THE STUDY OF HUMAN ATTACHMENT

People's experience of themselves as beings in relation has received many forms of expression. Psychologists' interest in this phenomenon is relatively recent. Though I will concentrate on psychological studies of attachment because they suggest ready connections to health care, this does not imply any denigration of other perspectives on human connectedness, particularly religious understandings. Religious thought and experience are at once a record of and commentary on people's experience of being-in-relation. Though I cannot do justice to the religious perspectives here, some reference to them, albeit allusive and selective, will place the psychological discussion in a richer context.

Major phenomenologists of religion (e.g., Otto [39], van der Leeuw [28]) agree that human religiousness expresses an encounter with a power that is experienced as wholly Other. At the same time, the Other is experienced as standing in some relation to oneself. To remain within the context of western Christianity and Judaism, "God" is understood as utterly transcending humanity, but also as Creator and Sustainer of the human world. Thus, Schleiermacher [47] writes of the feeling of "absolute dependence" as the central component of religious experience, and Paul's Letter to the Romans asserts that neither death, height, nor depth exceeds the reach of the love of God (8:38–9).

In the Jewish tradition, human being is unthinkable apart from connectedness to God and community. Abraham Heschel writes,

> Man in his being is derived from, attended by, and directed to the being of community. For man *to be* means *to be with* other human beings. His existence *is* coexistence. He can never attain fulfillment, or sense meaning, unless it is shared, unless it pertains to other human beings ([21], p. 45).

Moreover, Heschel asks,

> What are all prophetic utterances if not an expression of God's anxiety for man and His concern with man's integrity? A reminder of God's stake in human life; a reminder that there is no privacy? No one can conceal himself, no one can be out of His sight. He dwells with the Israel "in the midst of their uncleanness" (Leviticus 16:16). Living is not a private affair of the individual. Living is what man does with God's time, what man does with God's world ([20], p. 356).

Martin Buber, for whom the traditional language of faith in the living God is rendered problematic by the historical events and philosophical currents of the twentieth century, finds religious significance in the fact of human relatedness itself. Buber's account of human nature is rooted in the yearning and capacity for relationship. He gives this extended expression in *I and Thou*, where a recurrent theme is the nascent capacity for relationship in the human infant. Buber's language provides a striking transition to the psychological studies of early attachment behavior:

> Every child that is coming into being rests, like all life that is coming into being, in the womb of the great mother, the undivided primal world that precedes form. From her, too, we are separated, and enter into personal life, slipping free only in the dark hours to be close to her again; night by night this happens to the healthy man. But this separation does not occur suddenly and catastrophically like the separation from the bodily mother; time is granted to the child to exchange a spiritual connexion, that is, *relation*, for the natural connexion with the world that he gradually loses. . . .
>
> It is simply not the case that the child first perceives an object, then, as it were, puts himself in relation with it. But the effort to establish relation comes first – the hand of the child arched out so that what is over against him may nestle under it; second is the actual relation, a saying of *Thou* without words, in the state preceding the word-form; the thing, like the *I*, is produced late, arising after the original experiences have been split asunder and the connected partners separated. In the beginning is relation – as category of being, readiness, grasping form, mould for the soul; it is the *a priori* of relation, *the inborn Thou* ([8], pp. 25, 27).

Here – when the child first manifests the desire for connectedness and relation – the psychologists' studies of attachment behavior begin.

JOHN BOWLBY ON ATTACHMENT AND SEPARATION

In 1950 the World Health Organization asked the British psychoanalyst John Bowlby to study the mental health of homeless children. The preparation of his report intensified Bowlby's interest in the effects on children's personality development of separation from their mothers. Except for Anna Freud's reports on children removed from their London homes during the bombings of World War II [16], little systematic study had been made of this subject. Bowlby assembled a vast amount of data, supplementing his own clinical and research experience with extensive ethological research and a growing child development literature. By this time he had widened his study to the entire subject of the child's ties to the mother, and the effects on children when those ties break [5, 6, 7].

Bowlby's work is too rich in empirical detail and theoretical nuance to be adequately summarized here. By describing some of his main observations and conclusions, however, I hope to provide a basis for further discussion of the patient's experience of health care. I will highlight two themes: (1) human beings' predisposition toward intense attachment with nurturant figures in their environment; and (2) the characteristics of attachment figures that are most likely to meet people's needs.

Bowlby uses the term "attachment behavior" to cover a wide range of behaviors that are observed in the young of many animal species, including birds, sub-human primates, and humans [5]. These behaviors have as their common effect that the young maintain close physical proximity to the mother, either by seeking out and drawing near to the mother, or by eliciting a response on the part of the mother to come to the young and satisfy its needs. Examples include clinging, sucking, following, smiling, and – especially when separated – crying and expressing anger. Displays of these behaviors are typically intensified under conditions of alarm, pain, fatigue, or illness.

Bowlby asked, (1) What motivates attachment behavior? (2) What functions does it serve? (3) How do infants recognize and select their primary attachment figures? (4) What characteristics and behaviors of attachment figures most reliably satisfy the infant's needs?

To answer the question of motivation, Bowlby rejected an approach that was then dominant in the psychological literature, the theory of "secondary drive." According to this theory, behavior is motivated by a few basic needs, especially food, warmth, and sex. Since proximity to the mother is required for food, the infant is motivated to keep the mother close at hand. On this view, attachment behavior has no intrinsic motivation of its own, but is simply behavior that is reinforced when it repeatedly results in the satisfaction of a more basic need.

Although the theory of "secondary drive" appears widely in psychological explanations of attachment behavior, it does not fit empirical observations. If attachment behavior is primarily motivated by hunger, for example, why do infants persist in visual play, smiling and cooing, and other forms of contact even after they have been adequately fed? What explains observations by René Spitz [49], Anna Freud [16], and others, that children become angry, depressed, or withdrawn – despite total provision for their physiological needs – in the absence of affectionate, responsive, and familiar figures? If attachment behavior is the result of learning that is reinforced through the satisfaction of needs, why do monkeys prefer a soft and familiar cloth doll that provides no food to a wire mesh doll that will supply food when the monkey clings to it (the monkeys protesting vigorously when the cloth doll is removed and grasping it tightly when it is returned to them)? Why do monkeys, and many children, react to being pushed away by their principal attachment figure by clinging even tighter [5]?

Bowlby theorized that attachment behavior is a behavioral system with its own internal organization and its own function within the individual's motivational economy. The animals studied, including human infants, appear programmed to behave in ways that maintain proximity to others. And this behavior appears *independent* of the satisfaction of other, physiological needs. Attachment, then, is not merely the means to a more primary goal. In a fundamental sense attachment *is* the goal. Behavior aiming to achieve it persists long after the other goals have been accomplished and can be provoked, even to desperate degrees, when attachments are threatened in the midst of otherwise total environmental provision.[1]

As an ethologist, Bowlby is inclined to describe a behavior's function in terms of its survival value in the animal's "environment of evolutionary adaptedness." In the case of attachment behavior, he argues that for the sub-human primates as well as for humans, there was great danger in being separated from the group. Predators were more likely to attack an isolated individual; familiar individuals and territory seem to have a calming effect on many species (including humans), facilitating the execution of higher order tasks that also contribute to survival. (This may account for the observation that particular animals typically occupy a relatively small region within their ecological range.) Indeed, part of the process that induces attachment behavior is the experience of alarm or anxiety at the awareness of impending or actual separation from attachment figures [6].

In human beings' environment of evolutionary adaptedness, then, there was considerable survival value in proximity to the group. Natural selection would tend to favor individuals with the ability to elicit and maintain social connectedness. Among the traits most likely to be favored would be: (1) the experience of alarm or anxiety in the face of impending or actual separation, and (2) the display of behavior that promotes the reestablishment of relationships with familiar figures and environments, resulting in the reduction in "separation anxiety" and the return of normal biological and psychological functioning. Precisely these characteristics, Bowlby argues, describe the situation of the human infant and its abilities to maintain its ties to the mother.

Yet the infant's attachment behavior is only half the story. Attachment behavior is a mutual, interactive system that also depends on the attachment figure's response. To be realized in life, Buber's "inborn Thou" requires certain characteristics of the child's environment. When the child's hand is "arched out," who will grasp it in return?

CHARACTERISTICS OF ATTACHMENT FIGURES

When Bowlby turns to the implications of his research for the question of mental health, he offers the following succinct thesis: "Whether a child or adult is in a state of security, anxiety, or distress is determined in large part by the accessibility and responsiveness of his principal attachment figure" ([6], p. 23).[2]

A crucial question is how the terms "accessibility" and "responsiveness" are to be understood. Given that Bowlby's research grew out of his concern for homeless children, it is not surprising that he places great emphasis on the stable relationship between the child and the mother. But this raises some key questions: How stable must this relationship be in order to avoid psychological harm to the child? Can others beside the mother serve as the child's principal attachment figure? Do children incline toward single or multiple attachment figures? What about children who, either for reasons of emergency (e.g., the London blitz) or social planning (e.g., the Israeli kibbutz) are raised by several caretakers?

A great deal of empirical research has addressed these questions. Bowlby himself interprets this research to support what he calls "monotropy," that is, "there is a strong bias for attachment behavior to become directed mainly towards one particular person and for a child to become strongly possessive of that person" ([5], p. 308).[3]

Indeed, some studies do suggest that children raised by a group of caretakers (such as children in orphanages, nurseries, and the like) suffer more anxiety and interpersonal difficulties than children raised in conventional nuclear families [36, 53]. Other studies report little significant difference [27, 32]. One explanation for the lack of consistency is surely methodological. It is difficult to draw firm conclusions when studies vary widely in cultural context, sample size, and definitions of key variables.

One of the most intensively studied groups of children is those raised on the Israeli *kibbutzim* [23, 29, 38, 40, 44, 48]. Here the findings, which are remarkably consistent, provide qualified support for Bowlby's "monotropy" hypothesis. Having multiple caretakers does not by itself necessarily imply insecure attachment or subsequent psychological difficulties. Children can cope quite well with a group of caretakers *as long as they can develop a dependable relationship with some individual within the group*. What appears to mitigate the psychological impact of serial separations for kibbutz children is the consistent availability (albeit for the prescribed time in late afternoon known as "children's hour") of warm, emotionally gratifying interchange between the children and their parents.

A study by Schaffer and Emerson [46] further specifies some desirable characteristics of attachment figures. They found that a mother's readiness to respond to an infant's crying, and her willingness to initiate social interaction with the child, were the best predictors of a child's strong attachment. Moreover, among the secondary attachment figures available to these children, those who provided physical care but little social interaction were less likely to induce strong attachment than those who initiated interaction but provided no physical care. This last finding reinforces Bowlby's critique of the theory of secondary drive, demonstrating as it does the relative independence of attachment behavior from the satisfaction of physiological needs.

The discussion so far permits the following preliminary conclusions about human attachment: (1) the need for attachment is a genetic predisposition of the human species; (2) it is intensified in strange surroundings, and in situations of alarm, fatigue, illness or pain; (3) although multiple caretakers are capable of producing a state of relatively secure attachment, there is a bias in the infant for intense attachment to a single figure; (4) the characteristics of secure attachment figures include responsiveness to the infant's expressions of need, willingness to initiate emotionally satisfying social interaction, and continuity; (5) the development of strong attachment is independent of the provision of physical care, and may even be thwarted by competent physical care in the absence of meaningful social interaction.

Before returning to the world of health care, I will add one more psychological perspective. Winnicott's work on maternal-infant interaction suggests a further characteristic of attachment figures that will need to be considered when evaluating the social dimensions of the health care team.

"THE HOLDING ENVIRONMENT"

D. W. Winnicott came to the study of psychoanalysis by way of pediatrics. His roots are evident in his emphasis on the dynamics of personality development in the first few months of life. He extended psychoanalytic theory and practice into early phases of human development that his predecessors treated with far less sophistication. Through direct observation of children (something Freud seems to have done only rarely) and the reconstruction of infantile experience in the analyses of his older patients, Winnicott tried to describe how human beings experience their first relations with the world.

In characteristically colloquial style, Winnicott asserts that the prerequisite for the infant's healthy development is "good enough mothering," or, in less gender-specific terms, "satisfactory parental care" [56]. He divides parental care into three overlapping stages, of which the first, "holding," will be of chief concern here. As Winnicott himself defines it,

> The term "holding" is used here to denote not only the actual physical holding of the infant, but also the total environmental provision prior to the concept of *living with*. . . . The term "living with" implies object relationships, and the emergence of the infant from the state of being merged with the mother, or his perception of objects as external to the self ([56], pp. 43–44).

Winnicott suggests that physically holding the infant is a form of loving, and may even be the only way the mother can show the newborn infant her love. In its broader sense the "holding environment" also includes the mother's readiness to empathize with the infant in order to provide "a live adaptation to the infant's needs." In listing characteristics of a "good enough" holding environment, Winnicott writes:

> It meets physiological needs. Here physiology and psychology have not yet become distinct, or are only in the process of doing so; and it is reliable. But the environmental provision is not mechanically reliable. It is reliable in a way that implies the mother's empathy ([56], p. 48).

Thus far, Winnicott's concept of the holding environment is consistent with Bowlby's criteria for attachment figures: reliability, continuity, empathy, and the provision of physical care in the context of interpersonal warmth. But Winnicott introduces another aspect: within a secure holding environment the infant is able to tolerate the anxiety and disequilibrium that inevitably accompany separation and individuation.

Here the metaphor of holding enlarges to include not only physical holding but also the connotation of "containment." Each developmental stage exposes the infant to previously unknown aspects of the environment and to new challenges. The infant begins to encounter an external world of "not-me," and the reality of objects he or she cannot control. The infant also encounters new bodily sensations arising from physical needs and organ maturation. Secure holding means that the resulting feelings of strangeness, helplessness, or fear will not rage out of control or threaten the developing ego with annihilation. Instead, the infant's awareness of being held within an environment of safety permits the infant to tolerate these intense negative feelings en route to developing an enlarged capacity for dealing with the world. Without this feeling of safety, it is more likely that negative feelings – and the potentially growth-producing experiences that evoke them – will be avoided or denied, resulting in a more rigid and constricted personality in later life.

By facilitating tolerance for negative emotions such as helplessness and fear, the holding environment promotes two significant developmental achievements: the capacity for regression, and the capacity for autonomy. Though at first these achievements seem paradoxical, they are also closely related.

It is unfortunate that "regression" has acquired the connotation of the undesirable reappearance of immature behavior – the very opposite, it would seem, of "autonomy," with its connotations of maturity and independence. The behaviors that are usually at issue – dependency, the desire for nurturance, suspension of organized, rational thinking, and heightened emotionality – are actually supportive of many "adult" behaviors, and are quite adaptive in many situations. Psychoanalytic theorists have tried to rescue the term with expressions such as "regression in service of the ego," to indicate that in some circumstances (sexual intercourse is a prime example) highly valued experiences and relationships are enhanced when people allow themselves to feel out of control, vulnerable, and dependent.

Similarly, children's play often involves moving out from established boundaries of security. It carries the risk of reexperiencing feelings of frustration, helplessness, or anxiety associated with earlier periods of life. Yet play permits children and adults to test new cognitive or physical powers, a major step toward increasing competence and autonomy. The ability to play is thus intimately linked to the growth of maturity and independence, even as it requires lowering defenses against "childish" feelings or behavior.

Winnicott argues that both the capacity for regression and the development of autonomy are the fruits of a secure holding environment. He summarizes his view in an essay aptly titled, "The Capacity to be Alone." The basis of this capacity, he writes, "is the experience of being alone in the presence of someone." In other words, the infant discovers that it is safe to play, to explore, and to encounter strange objects and feelings, because someone is there should things begin to go out of control. Gradually, Winnicott continues, the confidence that someone is there is internalized, much as Bowlby describes the development of models of attachment figures who are predictably accessible and responsive. Then, even when the person is actually alone, he or she does not really *feel* alone:

> . . . there is always someone present, someone who is equated ultimately and unconsciously with the mother, the person who, in the early days and weeks, was temporarily identified with her infant, and for the time being was interested in nothing else but the care of her own infant ([56], p. 36).

WHAT DOES THE STUDY OF ATTACHMENT HAVE TO DO WITH HEALTH CARE?

I have been developing the idea that every aspect of human experience is conditioned by the social nature of human existence. I first presented evidence for the prominence of behavior that aims to maintain attachment to others. I then focused on the characteristics of attachment figures that most consistently satisfy human beings' need for connectedness. Those characteristics are:

(1) Dependable accessibility
(2) Empathy and responsiveness
(3) Readiness to initiate social interaction
(4) Continuity in the context of multiple subsidiary figures
(5) Willingness to permit regression and the expression of negative emotions

Before translating these characteristics into criteria for the health care team, I must anticipate questions that may have been troubling some people from the beginning. Does the attachment behavior of young children really have anything in common with the experience of illness and health care? Indeed, have I not, by concentrating on the child's relations with its parents, aggravated the tendency to infantilize the medical patient and treat him or her paternalistically, from which modern medical practice has only begun to recover? In short, does the foregoing discussion have anything to do with the matter at hand?

At least three lines of argument make it reasonable to suppose that people bring with them into health care settings the predispositions, needs, and expectations regarding attachment that are already manifest in the infant. These arguments, which I will sketch only briefly, suggest that illness and health care are permeated with attachment behavior, and – whether they are aware of it or not – health professionals inescapably play the role of attachment figure.

A. Stress and The Intensification of Attachment Behavior

The role of alarm, fatigue, illness, and pain in intensifying the need for attachment has already been noted in Bowlby's research. Precisely these factors motivate many medical consultations. Moreover, studies of adults in a variety of stressful situations, including illness, reveal a consistent pattern of heightened needs for affiliation with significant others, fears of abandonment and isolation, and the reactivation of childhood feelings of helplessness and dependence [22, 58]. As a social environment, the hospital is likely to increase these feelings [52].

To be sure, no unqualified generalizations are possible. Individuals vary in their responses to illness according to their personality styles, developmental histories, previous experiences of health care, and other factors. I would contend, however, that the burden of proof belongs on those who would deny the presence of intensified needs for attachment. Even people who at first appear aloof or fiercely independent, for example, frequently turn out to harbor strong desires for nurturance and support, which emerge when they are convinced that it is safe to lower their guard and acknowledge those needs [24].[4]

B. "Insecure Attachment" and the Decision to Seek Medical Aid

Considerable epidemiological evidence supports the connection between a person's loss of a support system and subsequent entry into the medical system [33]. People who have lost significant attachment figures, or are chronically uncertain as to their accessibility and responsiveness, are frequent users of health services. They are at higher risk for physical disease, frequently recover more slowly from disabilities, and have higher needs for attention and support from health professionals than people whose social networks are intact. For these patients, going to see the doctor is itself an example of attachment behavior.

Although the social and psychological motivations for seeking medical aid are frequently masked by the presentation of a physical complaint, skillful interviewing frequently elicits them. The patient's need to strengthen a sense of attachment and social connectedness emerges as a major issue – if not *the* issue – underlying the visit [2].

C. Regression and Transference in the Therapeutic Relationship

I suggested earlier that many situations in adult life call for the lowering of defenses against feelings of vulnerability and dependency, in the service of valued goals. Seeking medical care is one of those situations. To benefit from medical care a person must be able to accept help, which may conflict with feelings of pride or self-sufficiency. In psychotherapy, this is facilitated by what Phyllis Greenacre [18] calls an attitude of "expectant dependent receptiveness" toward the therapist. Greenacre argues that this attitude is derived from early experiences of parental care, and is adopted toward the therapist as part of the transference phenomenon. Writing in the context of general medicine, Eric Cassell [10] argues that "transference" is an expression of the desire for bonding with others at a time of need, and permits the relaxing of basic territorial defenses before the physician's probing questions and hands.

Providing medical care entails the creation of settings in which patients fell it is safe to experience two types of negative emotions. Emotions of the first type are directly associated with illness itself: uncertainty and anxiety about strange bodily sensations, fear of pain, frustration at disability, fear of death. Emotions of the second type arise from the social dimension of receiving care: anxiety in unfamiliar surroundings, the need to depend on strangers, and the need to accept help.

If the goal of medical care is to enhance autonomy – in decisionmaking and in daily living – then it is crucial to appreciate that genuine autonomy can only emerge from a matrix of dependency. The idea that autonomy is the polar opposite of dependency is conceptually shallow and psychologically bizarre. The gratification of a certain level of dependency needs provides the base of security and tolerance for anxiety required to extend one's sphere of competence and autonomy, as a developing child (a la Winnicott, above) or as an adult.

We have encountered these ideas before. They are central aspects of the holding environment. *Health care is the creation of a holding environment to help patients tolerate a variety of negative emotions, en route to regaining or enhancing their autonomous, independent function.* From this perspective, a crucial question to ask of any health care setting is whether it satisfies the conditions of a *safe* and *secure* holding environment. With this question in mind, we are ready to translate the criteria for secure attachment figures into patient-centered criteria for assessing health care teams.

PATIENT-CENTERED PERSPECTIVES ON THE HEALTH CARE TEAM

Without further ado, I propose that if the social structures of health care are to correspond to people's experience of illness, they must satisfy the following criteria:

(1) They will be organized and staffed according to the principle that the social dimension of health care is indivisible from the provision of physical care, and that each caregiver affects the patient's experience of being-in-relation.

(2) All caregivers, regardless of their technical specialty, will demonstrate interpersonal competence and awareness of the social dimensions of illness and health care, and will receive periodic evaluation of their performance in these areas.

(3) A single caregiver will be consistently available to the patient and will feel personal responsibility for the patient's well-being.

(4) All personal contacts with the patient will be characterized by empathy and responsiveness, based on familiarity with the patient's individual style of expressing needs.

(5) Patients will feel secure in asking for the help they need, and will be convinced of the acceptability of expressing negative emotions such as fear, anger, vulnerability, and helplessness.

These criteria are hardly new to the discussion of health care. What is surprising, however, is their almost total absence from evaluations of the health care *team*. Therefore, it remains an open question whether the team is a social structure that can reliably satisfy them.

What data exist warn that unless special effort is made, team care easily results in fragmentation and discontinuity in the patient's experience of being-in-relation. Strauss and his colleagues [51], for example, studied the effect of specialization and teamwork on several components of medical work in a large hospital. They concluded that problems of accountability, information transfer, and visibility – due to the complex division of labor – introduced significant gaps in interpersonal relations with patients. In one study of ambulatory care, an internist who employed nurses and physician assistants observed that patients would almost never leave the office without at least some personal interaction with him, even when all their "official" business had been thoroughly handled by ancillary personnel [9]. Physician retirement and the termination of physician-patient relationships at the end of residency programs can impose significant emotional strain on patients who must seek relationships with new caregivers [30, 54].

By contrast, it is commonly argued that teams adequately address the social dimension of health care because some team member is assigned to that "aspect" of illness. Yet the evaluation measures are so crude, and patient-centered variables so poorly represented in the empirical literature, that at present this claim must be seen primarily as wishful thinking.

Proponents of the team approach might be correct. Nor do I want to create a false dichotomy between technical and bureaucratic criteria and what I am calling patient-centered concerns. After all, patients are concerned about technical competence and cost when they select health care providers. My point is simply that evaluations of the health care team have been so dominated by technical, bureaucratic, and economic criteria that the evidence for the team's effectiveness as a patient-centered *social structure* has yet to be collected.

There is no doubt that technical and economic variables more easily lend themselves to quantification and comparison, which may partially account for their prominence in the literature. Therefore, one implication of my argument is the need for more imaginative social research to redress this balance.

Behind the methodological point, however, are two policy questions. The first concerns the locus of medical decisionmaking. Elsewhere in this volume, Raymond S. Duff describes a context for shared decisionmaking that he calls the "moral community." Implicit in his account, I believe, is a fabric of social relationships that satisfies the criteria that I have developed here. In other words, the criteria for a safe and secure *holding environment* are also the criteria for a *moral community* in which, as Duff puts it, "trust, education, and mutual respect are central values," and in which "physical, personal, and social conditions are given thoughtful consideration together in deciding and providing care." The failure to examine health care teams from this perspective, then, may well amount to the ratification of traditional hierarchical, authoritarian decisionmaking, by neglecting the social conditions required for a patient-centered alternative.

Second, what if the failure to explore patient-centered criteria as thoroughly as medicocentric ones expresses a consensus in policy circles that technical, professional, and economic concerns deserve priority in the scale of societal values? There is some evidence that these priorities may indeed reflect the values of patients who are *younger* and *generally healthy*, but not those of many elderly patients, or patients with multiple chronic illnesses. These people – who represent the fastest growing segment of the health care public – frequently opt for continuity of care and strong personal relationships when choices have to be made [14, 31].

The question of appropriate criteria for assessing health care teams reflects larger tensions in contemporary society: between personal and bureaucratic approaches to human needs; between the affective and technical aspects of professional work; and between an activist, youth-oriented culture and a growing number of elderly with different needs and values. Policies regarding the organization of health care will not be decided in isolation from these broader social currents. As choices are made, however, it will be important for patient-centered perspectives to guide research and to influence the debates.

College of Medicine
The Pennsylvania State University
Hershey, Pennsylvania

ACKNOWLEDGMENTS

I wish to thank Alice Wygant, of the Moody Medical Library at The University of Texas Medical Branch, for research assistance, and Patricia A. McRoberts and Alice Lundquist for preparation of the manuscript.

NOTES

[1] In concluding that the need for attachment is a primary feature of human development in its own right, Bowlby is consistent with other developments in psychoanalytic thought

in the 1940s and 1950s. His British countrymen, W. R. D. Fairbairn [13] and D. W. Winnicott [55, 56], had challenged Freud's position that sexuality is the primary force motivating infantile behavior. Standing a key tenet of Freudian theory on its head, these British "object relations theorists" argued that infants do not seek out contact with other people (or "objects" of attachment) in order to gratify sexual impulses, but that sexual behavior is one avenue for gratifying the desire to be in relationship with others.

[2] In a later passage, Bowlby elaborates this thesis in language that not only provides a theoretical context, but also calls to mind the situation of the anxious patient about to enter the health care system:

> In the working model of the world that anyone builds, a key feature is his notion of who his attachment figures are, where they may be found, and how they may be expected to respond. Similarly, in the working model of the self that anyone builds a key feature is his notion of how acceptable or unacceptable he himself is in the eyes of his attachment figures. On the structure of these complementary models are based that person's forecasts of how accessible and responsive his attachment figures are likely to be should he turn to them for support. And, in terms of the theory now advanced, it is on the structure of those models that depends, also, whether he feels confident that his attachment figures are in general readily available or whether he is more or less afraid that they will not be available – occasionally, frequently, or most of the time.
>
> Intimately linked to the type of forecast a person makes of the probable availability of his attachment figures, moreover, is his susceptibility to respond with fear whenever he meets any potentially alarming situation during the ordinary course of his life ([6], p. 203).

[3] This has led some people to criticize Bowlby for overemphasizing the importance of the mother as the unique and indispensable caretaker for the child, and for (unwittingly) perpetuating a cultural bias against women's spending more time out of the home [27, 45]. In fact, Bowlby's presentation is more complex, as are the findings that he and others have reported.

[4] Demographic and socioeconomic variables influence patients' preferences for attachment to individual health professionals. Significantly, older patients place greater weight on continuity of care (defined in one study as "seeing the same doctor for every visit") than on any other aspect of care [14]. This may be a highly relevant finding, given the increasing age of the health care population.

BIBLIOGRAPHY

1. Barnard, D.: 1985, 'Unsung Questions of Medical Ethics', *Social Science and Medicine* **21**, 243–249.
2. Barsky, A. J.: 1981, 'Hidden Reasons Some Patients Visit Doctors', *Annals of Internal Medicine* **94**, 492–498.
3. Bloom, J. R.: 1979, 'Team Care: Solution for Hospital Oncology Units?' *Health Care Management Review* **4**, 23–30.

4. Boss, M.: 1983, *Existential Foundations of Medicine and Psychology*, Jason Aronson, New York.
5. Bowlby, J.: 1969, *Attachment and Loss: Vol. 1: Attachment*, Basic Books, New York.
6. Bowlby, J.: 1973, *Attachment and Loss: Vol. 2: Separation*, Basic Books, New York.
7. Bowlby, J.: 1980, *Attachment and Loss: Vol. 3: Loss*, Basic Books, New York.
8. Buber, M.: 1958. *I and Thou*, 2nd ed., Scribner's, New York.
9. Burnum, J. F.: 1973, 'What One Internist Does in His Practice', *Annals of Internal Medicine* **78**, 437-444.
10. Cassell, E. J.: 1979, *The Healer's Art*, Penguin, Baltimore.
11. Dingwall, R.: 1976, *Aspects of Illness*, Martin Robertson, London.
12. Donabedian, A.: 1983, 'The Quality of Care in a Health Maintenance Organization: A Personal View', *Inquiry*, **20**, 218-222.
13. Fairbairn, W. R. D.: 1952, *Psychoanalytic Studies of the Personality*, Routledge and Kegan Paul, London.
14. Fletcher, R. H. *et al.*: 1983, 'Patients' Priorities for Medical Care', *Medical Care* **21**, 234–242.
15. Frank, J.: 1973, *Persuasion and Healing*, Johns Hopkins University Press, Baltimore.
16. Freud, A. and Burlingham, D.: 1973, *Infants without Families: Report on the Hampstead Nurseries 1939–1945*, International Universities Press, New York.
17. Gold, M.: 1976, 'A Crisis of Identity: The Case of Medical Sociology', *Journal of Health and Social Behavior* **18**, 160–168.
18. Greenacre, P.: 1954; 'The Role of Transference', *Journal of the American Psychoanalytic Association* **2**, 671-684.
19. Halstead, L.. 1976, 'Team Care in Chronic Illness: A Critical Review of the Literature of the Past 25 Years', *Archives of Physical Medicine and Rehabilitation* **57**, 507–511.
20. Heschel, A. J.: 1955, *God in Search of Man: A Philosophy of Judaism*, Farrar, Straus, Giroux, New York.
21. Heschel, A. J.: 1965, *Who is Man?*, Stanford University Press, Stanford, California.
22. Janis, I.: 1958, *Psychological Stress*, Wiley, New York.
23. Kaffman, M.: 1972, Characteristics of the Emotional Pathology of the Kibbutz Child', *American Journal of Orthopsychiatry* **42**, 692–709.
24. Kahana, R. J. and Bibring, G. L.: 1964, 'Personality Types in Medical Management', in N. Zinberg (ed.), *Psychiatry and Medical Practice in a General Hospital*, International Universities Press, New York, pp. 108–123.
25. Katz, J.: 1984, *The Silent World of Doctor and Patient*, Free Press, New York.
26. Kestenbaum, V. (ed.): 1982, *The Humanity of the Ill: Phenomenological Perspectives*, University of Tennessee Press, Knoxville.
27. Lamb, M. E. (ed.): 1982, *Nontraditional Families: Parenting and Child Development*, Erlbaum, Hillsdale, New Jersey.
28. Leeuw, G. van der: 1963, *Religion in Essence and Manifestation*, Harper and Row, New York.
29. Levy-Shiff, R.: 1983, 'Adaptation and Competence in Early Childhood: Communally Raised Kibbutz Children versus Family Raised Children in the City', *Child Development* **54**, 1606–1614.
30. Lichstein, P.: 1982, 'The Resident Leaves the Patient: Another Look at the Doctor-Patient Relationship', *Annals of Internal Medicine* **96**, 762–765.

31. Linn, M. W. *et al.*: 1982, 'Satisfaction with Ambulatory Care and Compliance in Older Patients', *Medical Care* **20**, 606-614.
32. Maccoby, E. and Feldman, S. S.: 1972, 'Mother-attachment and Stranger Reactions in the Third Year of Life', *Monographs of the Society for Research in Child Development* **37** (1, Serial No. 146).
33. McKinlay, J. B.: 1981, 'Social Network Influences on Morbid Episodes and the Career of Help Seeking', in L. Eisenberg and A. Kleinman (eds.), *The Relevance of Social Science for Medicine*, D. Reidel, Dordrecht, Holland, pp. 77–110.
34. McWhinney, I. R.: 1972, 'Beyond Diagnosis', *New England Journal of Medicine* **287**, 384–387.
35. Mechanic, D.: 1976, *The Growth of Bureaucratic Medicine: An Inquiry into the Dynamics of Patient Behavior and the Organization of Medical Care*, Wiley, New York.
36. Moore, T. W.: 1969, 'Effects on the Children', in S. Yudkin and A. Holmes (eds.), *Working Mothers and their Children*, 2nd ed., Sphere Books, London.
37. Nagi, S.: 1975, 'Teamwork in Health Care in the U.S.: A Sociological Perspective', *Milbank Memorial Fund Quarterly*, 75–91.
38. Nevo, B.: 1977, 'Personality Difference Between Kibbutz Born and City Born Adults', *Journal of Psychology* **96**, 303–308.
39. Otto, R.: 1923, *The Idea of the Holy*, Penguin, Harmondsworth.
40. Pelled, J.: 1964, 'On the Formation of Object-Relations and Identifications of the Kibbutz Child', *Israel Annals of Psychiatry and Related Disciplines* **2**, 144–161.
41. Pfifferling, J.: 1981, 'A Cultural Prescription for Medicocentrism', in L. Eisenberg and A. Kleinman (eds.) *The Relevance of Social Science for Medicine*, D. Reidel, Dordrecht, Holland, pp. 197–222.
42. Reiser, S. J.: 1978, *Medicine and the Reign of Technology*, Cambridge University Press, New York.
43. Rogers, J. and Curtis, P.: 1980, The Concept and Measurement of Continuity in Primary Care', *American Journal of Public Health* **70**, 122–127.
44. Sagi, A. *et al.*: 1985, 'Security of Infant-Mother, -Father, and -Metapelet Attachments Among Kibbutz-Reared Israeli Children', in I. Bretherton and E. Waters (eds.), *Growing Points in Attachment Theory and Research*, Society for Research in Child Development, pp. 257–275.
45. Scarr, S.: 1984, *Mother Care/Other Care*, Basic Books, New York.
46. Schaffer, H. R., and Emerson, P. E.: 1964, 'The Development of Social Attachments in Infancy', *Monographs of the Society for Research in Child Development* **29** (3), 1–77.
47. Schleiermacher, F.: 1958, *On Religion: Speeches to its Cultured Despisers*, Harper and Row, New York.
48. Spiro, M. E.: 1958, *Children of the Kibbutz*, Schocken, New York.
49. Spitz, R. A.: 1946, 'Anaclitic Depression', *Psychoanalytic Study of the Child* **2**, 313–342.
50. Stein, R. E. K. and Jessop, D. J.: 1984, 'Does Pediatric Home Care Make a Difference for Children with Chronic Illness? Findings from the Pediatric Ambulatory Care Treatment Study', *Pediatrics* **73**, 845–853.
51. Strauss, A. *et al.*: 1985, *Social Organization of Medical Work*, University of Chicago Press, Chicago.

52. Taylor, S. E.: 1979, 'Hospital Patient Behavior: Helplessness, Reactance, or Control?', *Journal of Social Issues* **35**, 156–184.
53. Tizard, J., and Tizard, B.: 1971, 'The Social Development of Two-year-old Children in Residential Nurseries', in H. R. Schaffer (ed.), *The Origins of Human Social Relations*, Academic Press, New York.
54. Toms, W. B.: 1977, 'An Analysis of the Impact of the Loss of a Primary Care Physician on a Patient Population', *Journal of Family Practice* **4**, 115–120.
55. Winnicott, D. W.: 1958, *Collected Papers: Through Paediatrics to Psychoanalysis*, Tavistock, London.
56. Winnicott, D. W.: 1965, *The Maturational Processes and the Facilitating Environment*, International Universities Press, New York.
57. Wise, H. *et al.*: 1974, *Making Health Teams Work*, Ballinger, Cambridge.
58. Wolfenstein, M.: 1957, *Disaster*, Free Press, New York.
59. Zimmer, J. G., *et al.*: 1985, 'A Randomized Controlled Study of a Home Health Care Team', *American Journal of Public Health* **75**, 134–141.

PETER J. MORRIS

WHO CHARTERED THIS SHIP?

Team medicine is not a recent invention. The reality of the physician as captain of the ship was eroding before bureaucratic-professional, technological-scientific, and economic arguments or justifications for teams existed. The physician as captain of the ship was endangered when the passengers – or patients – mutinied, not because of changes in the crew. The crew has certainly influenced how the ship sails, but this ship sails for its passengers.

Patients seek medical care for numerous reasons, which are not limited to relief from acute symptoms or complaints of a chronic illness. Whatever the wisdom of the medical provider, it is the patient who initiates the contact and determines whether to continue the relationship, seek care elsewhere, or drop the whole matter. In an era of impending physician surplus and multiplying allied health professionals, patient preferences are beginning to receive more attention.

Marketplace pressures are suggesting that the medical profession take note of the aphorism long recognized by other service industries: the customer is always right. Lessons from the marketplace also suggest that teamwork is in. The customer is better off if the entire team works together to serve his needs.

But is this true for the health care industry? Is the patient's interpretation of illness as valid as the professional's? Should professionals accede to patient demands for a greater participation in the healing process? Should professional authority be balanced by patient autonomy? Is patient involvement therapeutic? Is the sick role dead? Is a team approach to medicine in the best interest of the patient?

Barnard suggests that the patient's experience of illness is the best starting place to evaluate the performance of the medical team. Illness is experienced in a social context. The malaise felt by an individual is translated into words and feelings recognized as symptoms by a social group, usually the family. The social group often provides the definitive response to such symptoms, offering comfort and treatment outside of the formal medical system. Only 12% of symptoms recorded by patients and families result in a visit to a medical office [1].

Nancy M. P. King, Larry R. Churchill, and Alan W. Cross (eds.)
The Physician as Captain of the Ship: A Critical Reappraisal, 113–122.

Symptoms may also fall outside of the common experience, or may be generally accepted as beyond the scope of informal remedies. If symptoms are recognized as serious or threatening, or as having medical or social consequences; if the barriers to seeking treatment are few and not insurmountable; and if the treatment is anticipated to be feasible and efficacious; then a visit to the formal medical sector may occur [10].

This formal medical sector is not limited to the physician's office, but includes a number of responses and a variety of providers. These responses are not mutually exclusive, and the patient may seek treatment from more than one source. While the choice among these providers may seem inherently an individual one, it is seldom made in isolation. Age, sex, race, ethnicity, family and cultural traditions all influence this choice [9].

This process confirms that from the patient's perspective, medicine practiced by a group of providers is not new. Patients have long sought to influence their experience of illness by choosing among a variety of providers. A patient may try a home remedy, sample the treatments of a number of providers – physician, chiropractor or curandero – and offer prayer and sacrifice through a minister, all in the hopes of easing symptoms, curing a disease, or influencing the outcome of illness. Perhaps what is new is the team that is organized by its members, rather than called into play piecemeal by patients. Typical teams today are presented to patients as complete packages.

Kraybill's chronology of the rise of neonatology and the NICU team does not represent the typical evolution of a medical team. The description implies that teams have been created de novo in response to new medical demands, breakthroughs in science and technology, and the emergence of new health professionals. These pressures argue well for the team concept in neonatology, but team medicine is not limited to subspecialty areas. Lynaugh's history of the changing relationship of nurse and physician, of the tensions and compromises that have occurred in that most familiar of medical settings, the hospital, is a more apt description of the rise of team medicine [6]. The experience of team medicine is as common to the patient as the medical receptionist and office nurse of the solo practitioner, though few would call this arrangement a team. I suggest that it is a team, and that healing begins with the receptionist. If we are met with rudeness, barriers to effective care rise; if the nurse reacts casually to our symptoms, the seriousness of our ailment is called into question.

We are socialized to use health care; that is, our expectations about health care are shaped by both our immediate environment and the larger society. Family and generational influences on the use of care have been demonstrated [5, 8, 12]. These influences are based on the views and experiences of family members, which in turn are based on the caregiving models and traditions made available by society. In our society, we expect care to be rendered by a physician, or, more likely, a caricature of a physician, a traditional stereotype. This is particularly so when we only rarely need to seek care. Encounters limited to an emergency room, urgent care center, or private office still rely primarily on the physician as the principal provider of health care, though even these settings have seen the addition of allied health providers. However, a patient who has a chronic condition requiring frequent use of health care is less likely to be seen by a stereotype and more likely to be seen by a health care team. Like the neonatal intensive care team, this team will probably have been created to serve specific patients and conditions, as in a hematology-oncology clinic, where medical and clinical nurse specialists, physical and radiation therapists, social workers, and psychologists all participate regularly in patient care and become familiar faces over time.

I worked for several years at Frontier Nursing Service in rural Southeastern Kentucky. Founded in the 1920s by Mary Breckinridge, a legend in American nursing history, the Frontier Nursing Service is dominated by nurses: midwives, nurse practitioners, and nurses in more traditional nursing roles. A physician there is interviewed to see if he or she is compatible with the nurses, not the other way around. Physicians are outnumbered by advanced level nurses, and all have equal votes on the medical staff. There, I was socialized to work with nurse practitioners. The agenda for what we called joint practice had been worked out by nurses, and we shared a collegial relationship not unlike those of partners in a group practice.

The system is driven by mutual respect for interpersonal and medical or nursing skills, and a belief in saying, as often as necessary, three simple words: "I don't know." Physicians and nurse practitioners are driven both to provide the best care for their patients and to ask for help when they need it.

Care at the Frontier Nursing Service is dictated by the special needs of its patients, whether or not those needs are primarily medical, and is tailored to a unique geographic, economic, and social subculture. The expertise of these team members is not limited to those skills imparted by training or education. In many ways, my personal skills made me a better "nurse" than some of the nurses on the staff, and their diagnostic skills made them better "physicians."

Certainly, there are economic arguments for the Frontier Nursing Service teams. Nurse practitioners have been shown to provide quality care at a cost-effective care price [11]. There are technical-scientific rationales, too, as protocols of team care are designed to move patients to the level of expertise they need to manage their problems. The professional-bureaucratic rationale is turned upside down in this nurse-dominated system, but exists as well. Mary Breckinridge started this experiment in rural care for the underserved people of Leslie County. She wished to demonstrate the utility of the nurse midwife in addressing maternal and child health needs. Perhaps unwittingly, she also contributed to a unique type of socialization of this population. Having rarely seen physicians, the population came to expect being seen by a nurse. Patient acceptance is high. Nurses have not only the authority of medical protocols and state licensure, but the authority bestowed by patients, who come to the clinic in search of "their nurse."

As care has become more complicated, and certainly as cure has eluded us for many of the chronic diseases, the care and cure dichotomy between nurses and physicians has become outdated. We all must care. And, as part of this cure, patients should more and more be involved in the non-medical aspects of care decisions, as alluded to by Maulitz and Areen [2, 7]. Paternalism should give way to partnership, not only because it is right and long overdue, but because it is part of the therapeutic relationship.

Barnard cites studies of human attachment pertaining to the parent-child relationship, or, in a more general sense, the caretaker-dependent relationship. This research speaks to the patient's needs for attachment, for a "holding environment," where the patient experiences relief and safety. Society, though, wants the sick to earn this place of security by playing by a set of rules requiring the sick person to want to be well, to seek help, and to cooperate with efforts to become well. Because of illness, patients are relieved of usual societal responsibilities, but can continue in this sick role only as long as they seek to get better.

What is the patient's role in the traditional medical encounter? In both acute and chronic care settings, the patient is expected to "comply," that is, to follow the instructions of the provider and thereby improve his condition. The patient thus becomes "entitled" to the safety of the holding environment.

The concept of compliance has come under attack by patients and professionals alike. Compliance has come to imply a passive process, where a patient follows a prescribed set of instructions, more like a footsoldier carrying out orders than a tactician devising a battle plan. The advantages of involving the patient actively in the healing process have been described in the writings of Cousins, from the patient's point of view, and Knowles, from the physician's [3, 4].

Today's major causes of mortality – cardiovascular disease, strokes, accidents, and cancer – are largely the result of lifestyle decisions made by patients. A large number of the acute illnesses for which patients most often seek care have no cure, though relief may be offered from some symptoms. Chronic illnesses, by their very nature, are not cured but rather endured, though often ameliorated by treatment. Patient involvement in treatment offers the best chance for modification of lifestyle, alteration of habits, and adjustment to disability, all of which can decrease morbidity.

More patient involvement in the choice of therapeutic measures, and more concern for the "routine" yet unacceptable side effects of treatment, can greatly enhance "compliance." Medicine has accomplished much with its attention to biomedical cures, and this has required the professional-as-scientist role. The changing morbidity and lack of cure for diseases today suggest that a patient-professional partnership is necessary to care properly where cure cannot be accomplished.

What of the professionals? Can physicians go it alone? Do they have the necessary education and socialization to serve the patient as members of a team? How we teach may be as important as what we teach. What conditions characterize interactions between professionals – courtesy, respect, and interest? Or contempt, disdain, and anger? Do we need team medicine in order to form a partnership with patients? And is the patient considered a member of the team?

Given the increasing complexity of medical interventions, the rise of new health professionals was inevitable. Physicians cannot know it all, nor do it all. It takes a team to manage the technology now applied to patient care, though teams exist for reasons other than technology. The acutely ill neonate in an NICU is cared for by all of the personnel Kraybill has described – physician, nurse, respiratory therapist, social worker – but after the acute phase of illness has passed, and often long beyond discharge, another team of professionals is often involved in continuing care for the patient. The neurologist, developmental physical therapist, behavioral psychologist, and special education teacher all may be involved with the child and family for years to come. These providers rely less on technology than on touching, feeling human interventions to continue care.

More concern should be given to the process of educating health care professionals. As Kraybill notes, there are many layers of expertise available in the neonatal intensive care unit, even within a single discipline. We are good at improving expertise. But expertise should be acquired in personal as well as technical skills. As we train our physicians in the differential use and application of new life-saving and life-sustaining technologies, we need to remind them of the gift of touch. Neither care nor cure is given to the electronic monitor or lab printout to which all eyes are often glued.

Nor is the care of patients automatically ensured because a team has been created. Team medicine serves its members as well as its patients, and that confounds the team's analysis of its work. Who determined the needs the team would serve? Who decided on the structure? And, as Barnard asks, what criteria are used to determine the success of the team?

The practice of medicine has its own social context; but the medicalization of team health care may obscure that reality in a false objectivity. The formal organization of allied health providers into a health care team often presumes that such a team represents all the elements needed to care for the patient effectively. This assumption is presumptuous. As Barnard suggests, a team cannot effect a cure without recognizing the broader social context of a patient's illness.

The organization of health care teams might be considered accidental, planned, or evolutionary. There is a spectrum of organizational structures – vertical, horizontal, and matrices – as well as a variety of styles of leadership – monarchists and anarchists, autocracies and democracies. Some team organizations are formally codified in written policies and procedures; some change from day to day, confusing staff and patient alike. Aside from the formal structure described by an organizational chart, an informal structure exists, too.

Social tradition has dictated the structure of many of these teams. Where nursing has traditionally been strong, as in the private non-profit hospital or at Frontier Nursing Service, nurses are high in the hierarchy. In private offices, multi-specialty groups, or proprietary hospitals, physicians dominate. Administrators may be of more importance where regulations rule, as in some HMOs and insurance arrangements. Where innovation has a tradition, often the case of academic institutions, knowledge and expertise may dictate structure – as in the NICU described by Kraybill. In the private, for-profit sector, economics may be the driving force behind innovation.

More important than rank within a hierarchy are questions of order and responsibility. There must be a mechanism for decisionmaking. Should all votes be equal? Should expertise or tradition rule? Whatever the organization of the team, there are legal constraints which place the physician in the place of greatest – though not sole – liability, even as more and more allied health professionals carry malpractice insurance of their own.

Teams can be autocratic, democratic, or socialistic, implying unitary control, rule by majority, or to and from each according to needs and talents. Whatever the decision process, teams cannot be rigid and inflexible, because patients and their needs are not rigid and inflexible. Teams should not become compartmentalized. Patients' problems do not come in neat packets and neither should the services of a team. Just as every medical problem does not need a specialist, every social problem does not need a social worker. Though social workers might handle a problem best, they might also guide an intern to the solution. The presence or availability of a specialist should not be an invitation for the generalist to abdicate responsibility and call for a consult.

Kraybill's observations on the NICU give reason for concern. Are the various allied health specialties, from physical therapist to respiratory therapist to social worker, more interested in carving their part of the pie than in meeting a need of the patient? Have the needs of profession and livelihood superseded patient needs? Competent allied health professionals have often carved their niches in spite of, rather than because of, needs recognized by the medical profession itself. Physicians have not, in general, welcomed the new health professionals. In an era of increasing competition and a physician glut, licensing has been restrictive in many states for many reasons not limited to economics. Seen in its worst light, a selfish medical profession is jealously guarding its position in society and the marketplace. But, too, there is the question raised by Kraybill: Who *should* head the team? And there are other considerations as well: Who will police the medical marketplace? Who will provide quality assurance? And who will safeguard the rights of the patient? These are areas of legitimate authority for the physician. Barnard's criteria for the evaluation of the team fail to mention the problem of authority.

As Kraybill suggests, physicians are socialized for authority, responsibility, and decisionmaking. These qualities, and the force of tradition, have placed the physician in position as captain of the ship. The traditions are changing, and so are the structures of control. New structures must address the questions of how authority, responsibility, and decisionmaking may be shared. In an era of lifestyle-related diseases, these are traits we would like to see in our patients. The patient should not be overlooked as a potential head of the team, often its most knowledgeable member.

The true patient-centered perspective must realize that today's patients often have the education and socialization to choose a provider who can best serve their needs. In the infant's case, the parent or guardian often helps tailor the care to the problem. The parent becomes a professional consumer of health care. The recovering stroke victim is also such a professional, seeing the services of the speech therapist or the physical therapist as more important, at times, than the services of the neurologist. These people know the benefits of team medicine and know how to use the team for their benefit – showing themselves often to be better educated regarding the use of the team than the professional.

If a team is organized vertically, with someone on top, there must be someone at the bottom. Up to now, that someone has usually been the patient. A team organized horizontally is not necessarily one without leadership or direction. The needs, and indeed the desires, of the patient might best direct which professional should take the lead, a lead that may change hands as the patient's condition, or the team's interactions, evolve. Indeed, the patient with time and expertise to interview and choose among alternatives might be considered in the best position to head the team. Even as we recognize the difficulties that trained professionals encounter in choosing among competing claims and therapies, more efforts should be made to empower patients and to place them at the top of the organizational chart.

In all aspects of our society, not just medicine, we are choosing whether control and authority will be given to highly trained technocrats, or will be retained by less sophisticated persons, who might just have more common sense and perspective than the experts. If physicians, or teams, persist in becoming a more technocratic elite, they may find their legitimate authority eroded, and eventually taken from them. Expertise, offered arrogantly and without explanation, does not serve patients well. Neglect of this issue will, rightly or wrongly, lead to regulation or oversight.

Accountability, information transfer, and visibility can be enhanced or obscured by the team approach. It takes an awareness not only of the socialization of patients, but of the socialization of team members. Socialization of patients and providers is a constant and renewing process, one in which patients and providers alike must participate.

There will be tensions created by the process of empowering patients, and giving them their rightful place on the team. The role of team leader does not imbue the patient with the knowledge and expertise of a physician, nurse practitioner, physical therapist, or social worker. One would hope, though, that the informed team leader claims not expertise, but the right to choose, to question, and to adapt the offered advice. Patients may not choose wisely. In instances where there is reason to doubt the patient's inherent capacity to choose, there already exist ethical and legal approaches to resolve the quandary. The assignment of responsibility for an adverse outcome would need to be firmly delineated, to satisfy our litigious society. If patients are to be given a voice in authority, they must accept a share of accountability. Non-physician professionals already recognize their responsibility to the patient and accountability to physicians as dictated by regulatory boards, and have accepted liability for their actions. Professionals, whatever their field of expertise, will continue to maintain professional standards and be accountable to licensing authorities, though these policing boards need to improve the range and scope of their review and penalty processes.

Not enough attention has been given to these processes. Academic medical centers have long been responsible for experimentation in new and exciting therapies and approaches to disease, but approaches to disease are not always worked out in the laboratory. The wider medical community will need to be canvassed and consulted. Ethical and legal discussions will be necessary, not only among the academics, but in the streets. The patient's perspective must be considered, as must the process of transferring skills to the next generation of physicians, nurses, and therapists. If we neglect this process, the next generation of patients will pay the price.

Wake County Department of Health
Raleigh, North Carolina

BIBLIOGRAPHY

1. Alpert, J. J. *et al.*: 'A Month of Illness and Health Care Among Low Income Families', *Public Health Reports* **82**, 115–152.
2. Areen, J.: 1988, 'Legal Intrusions on Physician Independence', in this volume, pp. 39–65.
3. Cousins, N.: 1979, *Anatomy of an Illness*, W. W. Norton and Co., Inc., New York.
4. Knowles, J.: 1977, 'The Responsibility of the Individual', in Knowles, J. (ed.), *Doing Better and Feeling Worse: Health in the United States*, Norton, New York, pp. 57–80.
5. Littman, T. J.: 1971, 'Health Care and the Family: A Three Generational Analysis', *Medical Care* **9**, 67–81.
6. Lynaugh, J.: 1988, 'Narrow Passageways: Nurses and Physicians in Conflict and Concert Since 1875', in this volume, pp. 23–37.
7. Maulitz, R.: 1988, 'The Physician and Authority: A Historical Appraisal', in this volume, pp. 3–21.
8. Miller, F. J. S.: 'The Epidemiological Approach to the Family as a Unit in Health Statistics and the Measurement of Community Health', *Social Science and Medicine* **8**, 479–482.
9. McKinlay, J. B.: 1972, 'Some Approaches and Problems in the Study and Use of Services – An Overview', *Journal of Health and Social Behavior* **13**, 115–152.
10. Rosenstock, I. M.: 1966, 'Why People Use Health Services', *Milbank Memorial Fund Quarterly* **44**, 94–124.
11. Salkever, D. S. *et al.*: 'Episode Based Efficiency Comparisons for Physicians and Nurse Practitioners', *Medical Care* **20**, 143–153..
12. Tyroler, H. A. *et al.*: 1965, 'Patterns of Preventive Health Behavior in Populations', *Journal of Health and Human Behavior* **6**, 128.

SECTION III

TECHNOLOGY AND FINANCING: CHANGING THE COURSE

ROBERT M. COOK-DEEGAN

THE PHYSICIAN AND TECHNOLOGICAL CHANGE

Physicians derive their authority from at least two sources. Their moral authority is based on the obligation to place patient benefit above self-interest, which is reflected in professional codes of ethics. Technical authority arises from the physician's superior knowledge about the course of disease and means to prevent, reverse, or arrest it. In the physician-patient relationship, the basis for moral authority is a patient's trust of the physician. In health policy, moral authority hinges on public trust of physicians' aggregate motivations (i.e., are physicians interested in service or money?). These bases for moral authority have until recently been widely presumed to be healthy and stable.

The rapid disproportionate growth of technical authority has, however, led to questioning of this presumption. The revolution in medical technology has had important moral consequences. It has dramatically expanded the ability of physicians to benefit their patients medically. At the same time, however, it has also made care expensive, which in turn has increased the importance of economic incentives. Economic incentives in the United States often pit physician fiscal self-interest against the medical needs of patients. The conflict between economic benefit for physicians and the medical needs of patients is intensifying as more physicians are trained, more technologies become available, and pressures to reduce overall health spending increase.

The way technologies are developed and paid for in the United States directly contributes to the conflict, although it need not if incentives are restructured. Economic conflicts between physicians and patients may arise no matter what the setting: in for-profit and not-for-profit institutions, or in corporate organization and its alternatives. Different institutional organizations may exacerbate the conflict, but they do not create it. The conflict is grounded not in how care is delivered but in how physicians are paid for their services. The physician is unlikely long to be trusted who financially benefits from either excessively using technologies (e.g., in fee-for-service payment for physician services or outpatient use of technologies) or withholding care to boost either his own fiscal interest or that of his employer.

Nancy M. P. King, Larry R. Churchill, and Alan W. Cross (eds.)
The Physician as Captain of the Ship: A Critical Reappraisal, 125–158.

The ways that technological medicine is economically organized thus place the two sources of authority in conflict when physicians deal with patients. (Note that other situations that address very similar issues do not raise the conflict: e.g., choices of where to practice or what specialty to choose.) The physician's unquestionably superior grasp of the technical aspects of care is rising rapidly, but distrust of physicians is undercutting their moral authority. The way technology is developed and used has made this conflict more apparent.

CONSEQUENCES OF TECHNOLOGICAL GROWTH

Medical technology has vastly increased in power and scope over the last two decades. This explosion in the number of tools available to promote health has changed the ways physicians practice medicine. The primary effect has been to enhance what doctors can do and the benefit that patients can expect. The emphasis of this paper will be on the *secondary*, or indirect, effects of expanding technical capabilities in medicine. Current trends suggest that freedom of choice among physicians may be reduced by economic factors, government attempts to restrain health spending, and third party payers' attempts to constrain cost growth. The use of technology is one area where physicians reign supreme on a daily basis. In the long term, however, physicians engaged in direct patient care are likely to exercise progressively less influence about which technologies are developed and how their use is reimbursed.

Physicians now make most decisions about which diagnostic tests are used, which drugs prescribed, and which surgeries performed. In recent years, in part due to bioethical analysis, patients have played an increasing role in decisions, particularly when these involve substantial nonmedical considerations (e.g., when to terminate treatment near the end of life) or when the consequences involve significant danger to the patient (surgery and invasive diagnostic tests). Some physicians regard this as a loss of their autonomy. It is, however, a restoration of autonomy to its proper place – the patient. Technology has not been the overriding force in enhancing patient autonomy, but it has certainly increased the number of major decisions that need to be made by patients in consultation with their physicians. Technology has, therefore, forced health professionals and ethical analysts to confront difficult choices with increasing frequency.

Technical advance is likely to have several secondary consequences.

They are listed here, and the arguments supporting the predictions follow later in the text. Some of the important changes likely to occur in the next decade are:

- Diminished power of individual patients and physicians to determine the overall structure of health care or to control local availability of particular technologies,
- Increased power of government, other third-party payers, and large organizations (e.g., corporations, hospital networks, HMOs, etc.) to determine which technologies are used,
- Increased capital needs for development of new technologies, leading to more "commodification" of health care,
- Increased role of private firms in determining which technologies are developed,
- Increased specialization of physicians into disparate groups with diminished political cohesion, economic alignment, and ethical consensus,
- Diminished physician choice of career paths, geographic location, and type of practice,
- Increased regulation of new technologies and uses of existing technologies,
- Increased opportunities for corrosion of ethical values and moral authority of physicians as physicians come to be dominated by larger economic interests, and
- Increased role of government in aligning economic incentives with ethical values.

Many of these predictions are not based solely, or even primarily, on development of technology. The advance of technology, however, will contribute to each.

THE ROLE OF TECHNOLOGY IN MEDICAL PRACTICE

It would be an interesting exercise to reincarnate Sir William Osler and observe his reactions to modern medicine. His opinions cannot be known, but a few substituted judgments are unlikely to be wrong. His first reaction would probably be awe at the technological power that is now commonplace. Medical practices have changed dramatically, and the major killing diseases of his era are no longer the staples of daily medical care. Many diseases once rare – myocardial infarction and lung cancer, for example – have replaced tuberculosis, syphilis, smallpox, and polio. Walking into a modern hospital, Osler would doubtless be struck by the computers, diagnostic imaging devices, laboratory tests, and complex patient monitoring instruments. It is quite plausible, in fact, that the plethora of exotic technologies would overwhelm Osler's senses for the first few hours.

When he regained his bearings, he would probably note the advanced age of the patients – almost twice as old on average as those he cared for – again traced to technology (not entirely *medical* technology, but also application of knowledge about sanitation, workplace safety, and environmental hazards). We cannot guess what he would think about how medicine is organized, financed, or taught. The point is made nonetheless: technological advance is a main theme in modern medicine, perhaps the dominant one. Medicine's ethical precepts have changed much less than the technology of practice, and technology has forced on us more difficult choices [17]. (One might even argue that the increasing struggle with ethical dilemmas resulting from new technology is the main force behind the growth of bioethical analysis as a discipline. The new interest in careful ethical justification in medicine may derive from continued contact with moral conflict, exacerbated by moral pluralism.)

Many gains in life expectancy early in the century stemmed from reducing mortality among young and middle-aged individuals. Life expectancy changes in the last decade have, in contrast, been concentrated in the oldest segments of the population. To caricature these developments: those saved from polio are now treated successfully for hypertension and live into their eighties to die with Alzheimer's disease.

The extent to which this increase in life expectancy has resulted from medical technology is a topic of considerable debate. The definition of medical technology itself is open to debate, but loosely defined it refers to the tools and procedures used to improve health. In daily practice, it matters little whether it is environment, behavior, or medical technology that lengthens life – the choices at hand are usually restricted to using particular technologies or influencing the patient's behavior. The considerations in most medical decisions are local and narrow, based largely on assessments of medical risks and potential benefits rather than social or fiduciary implications. Aggregate changes in life expectancy result from a myriad of individual choices in medical care and daily life. Decisions about medical care are frequently made with fragmented and incomplete information. Well tested and thoroughly understood medical practice is the exception rather than the rule. Most routine medical decisions are based largely on shifting facts and a relatively unreliable (and often uninspected) web of beliefs held by individual physicians and patients.

Wide geographic variations in uses of most technologies attest to the uncertainties involved. Some procedures are 26 times as common in some regions as others (injection of hemorrhoids), while other procedures vary little among regions (lens extractions and inguinal hernia repair) [3]. The inability to explain such variations merely confirms the multiplicity of factors involved in decisions about use, the complexity of interactions among the factors, and the impoverished state of knowledge about optimal care [3, 8, 33]. The data on geographic variations have been interpreted to show that physicians retain considerable discretion in deciding which technologies to use. Yet the data do not indicate that physicians actually control the resources available to them. It seems more likely that a combination of local politics, administrative practices, and medical standards underlies the geographic variations. There is little doubt that physicians dominate medical markets in many areas. There is equally little doubt that individual physicians are increasingly dependent on large external sources of capital (e.g., for-profit and non-profit hospital corporations), and exercise progressively less individual control over their careers and major investment decisions that affect their practices.

Uncertainty about technology does not preclude its use. One of the skills acquired early in medical education is how to feel emotionally comfortable when making decisions in the absence of any reliable information. (This skill is often overextended, resulting in an inability to recognize what is not known, a common source of dogmatism in clinical medicine.[2]) The overwhelming uncertainties can mask an important core of reliable knowledge about how to use certain technologies to combat some diseases. One can well understand the laudable impact of treating hypertension with diuretics and beta-blockers, or removing the appendix in response to appendicitis. The effective control of bacterial microbes (through both antibiotic use and advanced sanitation) and prevention of some infectious diseases through immunization are other clear examples of medical prowess. This core of reliable information is what distinguishes modern medicine from its antecedents and legitimates the power of the healing professions.

History may repeat itself. Today's major killers – cardiac and vascular disease, stroke, and cancer – are yielding to new diagnostic and therapeutic approaches. If one removes the large cohort of patients suffering from a particularly intractable form of cancer – lung carcinoma due to smoking – then each of these modern diseases is either stable or declining in mortality ([22], Appendix A). It seems likely that we are in transition to a period in which most people will reach what used to be called old age in relatively good health. Today's killing diseases will probably remain the major killers but their onset will be delayed, and superimposed on a population afflicted with dementia, poor hearing, failing sight, and rotting joints. Data do not support Fries's idea of a "compression of morbidity," with multiple organ systems failing all at once at the end of the "natural" lifespan [9]. The failure of different organs does occur, but not suddenly. Rather, the modern octogenarian typically undergoes a protracted and progressive involution terminating in death. The period of dying may not be shortening, but rather lengthening [30], resulting in an "extension of morbidity."

Physicians are not putting themselves out of business, but rather generating more ill health with each notch of the technological ratchet. Ironically, this is generally to the better, because morbidity is associated with longevity. Apparently, most of us prefer to live long with arthritis rather than die young (and in otherwise good health) from infectious disease. The world is a better place for the presence of modern medical technology, but the blessing is not unqualified. Increased morbidity is not the primary or intended effect of medical technology, but nevertheless results from it. Reduction in infant mortality yields more people reaching middle age. Success in thwarting death in middle age (from stroke, cardiovascular disease, or cancer) results in a larger and older population suffering from multiple disabilities. Doctors have more to do.

The success of medical care has caused an escalation of costs stemming from several factors. In general, it costs more to do more. New technologies are increasingly costly and capital intensive. New knowledge has spawned increased *numbers* of useful technologies. More people suffer from ill health for *longer* periods. And the impressive increase in biomedical research, supported by Federal funding that started to grow rapidly after World War II, has generated a need for involvement of *more decisionmakers* with specialized knowledge.

The model of health care that characterized Osler's period was a physician-patient relationship in which physicians were dominant. Physicians, even at teaching hospitals, were primarily clinicians whose ethical stance was to judge medical benefit for their patients and act to promote it. The key act was making accurate judgments about patient benefit. Most medical decisions were made by the physician, and informed consent was not a well-developed doctrine.

This model has slowly given way, at least in bioethical analysis, to one in which the physician is not the captain of the ship, but rather a navigator, assisting the patient captain [10]. The key act here is imparting information in a useful and accurate form to the patient. The ultimate locus of difficult decisions resides in the informed consent of the patient. While it is debatable whether this model has ever been implemented, the ethical analysis of the last several decades clearly prefers it in most cases [6, 15]. Informed consent is likely to remain the dominant legal doctrine in the future, and one can hope that medical practice will adapt to the legal precepts in response to accumulating case law. (This is not inevitable, however, because the standards for informed consent determine the relative standing of patients and physicians, and these standards vary from court to court.)

The ascendancy of the ideal of patient-centered decisionmaking coincided with the first explosion of medical knowledge following World War II. Physicians took on more responsibility for creating new knowledge, and the beginning of fragmentation in roles began, separating scientist from clinician. Yet most physician investigators kept a hand in clinical practice, and a substantial fraction of each medical school class went into research. There was frequent switching between research and clinical care in medical careers. Physician autonomy during this period was less diminished than laterally displaced. Direct control in clinical situations was shifted more to patients, but there were other realms in which the physician still reigned supreme, such as the laboratory and the operating room.

This period is now ending, giving way to an era in which physician autonomy is being diminished in many areas. Basic research is so specialized that it is increasingly dominated by Ph.D. professional investigators. Clinical research is dominated by academic physicians relatively isolated from primary care. Now fewer physicians pursue careers in research.[3] Clinical decisions are increasingly constrained by administrative edicts, Professional Review Organization reviews, precepts for avoiding exposure to malpractice claims, third party standards for payment, criteria for government program eligibility, and hospital committee protocols for institutional practices (e.g., ethics committees, business committees, legal counsel). Even in the operating room, the last bastion of physician preeminence, there is a growing dependence on others in managing complex technologies (e.g., lasers, evoked potential recorders, prosthetic implants). In research the investigator must now answer to many entities: Institutional Review Boards, Institutional Biosafety Committees, Animal Care Committees, and government regulatory agencies.

Within the ranks of physicians, there is further division. Academic medicine is so esoteric and complex that it is increasingly removed from routine clinical practice. A cadre of specialists familiar with a small range of illnesses (although still strained by the effort to keep abreast of new developments regarding them) have a different set of clinical norms than "primary care" physicians who must manage the whole range of illnesses as patients walk in the door. The primary care physicians rely on "bridging" specialists (those who emerge from the academic labyrinth to give directions) who translate scientific and clinical research into practical rules for daily medical care. Academic medicine now resembles a fiefdom of intellectual city-states linked to but distinct from practitioners whose practices more closely resemble the traditional norm. The emergence of subgroups diminishes the unifying strength of shared values, especially as prestige within the academic elite is conditioned on publication much more than clinical perspicacity. The bridging specialists are highly valued physicians, usually senior enough to have survived the rigors of academic competition while miraculously maintaining their common sense.

The diminished power and increased fractiousness of physicians is intensified by the emerging glut of physicians, the trend for new physicians to accept salary arrangements with an employer (in contrast to the financial independence of solo or even group practice), and the bureaucratic encrustation of modern health institutions. Bureaucracy is consuming an increasing share of the health care dollar, particularly in the complex reimbursement environment of the United States [12]. This increased consumption of resources by administrative cost in turn implies a larger role for administrators in determining the nature of medical practice; the money they control reflects their power.

What physicians lose, patients are unlikely to gain; diminished physician independence seems unlikely to yield enhanced patient autonomy. While a diminished degree of physician dominance is necessary to promote patient autonomy, the *way* that physicians are currently losing power portends even more marked subordination of patient autonomy. Growing medical knowledge and technical complexity make informed consent more complex, even where it is legitimately sought. (It would be ironic indeed if patients' newly won authority in making medical decisions were squandered in a morass of technical bewilderment.) The policies of payers may supersede considerations of benefit to individual patients. The growth of health administration may encroach on patient choice, just as it impinges on the decisions of physicians. Most of these changes can be directly traced to the advance of medical technology.

The primacy of technology in rendering care more complex is not open to question; whether technology alone has caused the shift in physician and patient autonomy is, however, debatable. Another factor – the economic environment – also determines the way that medicine is practiced. Attempts to assess the relative contributions of economic versus technological factors on the growth of technical capacity and cost are futile. Economic factors could not have increased the size of the medical component of the economy if there had been no new technologies to enable the recycling of capital, and the technologies could not have developed in the absence of a fertile economic environment. Technologies and economic interests have fed upon one another to foster the growth of modern medicine.

Technology enhances the power to improve a patient's health, but the secondary effects of proliferating technology diminish the ability of any one individual to control its use. The number of choices has increased dramatically as a consequence of technological improvements, but the *proportion* of such choices that can be controlled by an individual patient or physician is shrinking.

Physicians have ceded authority to patients on ethical grounds (and under legal pressure) [6]; they have lost scientific and technical power to other professionals; and they are increasingly constrained by the dictates of administrators and payers in a capital-intensive and complicated economic environment.

Patients have gained ground on physicians in making decisions about their care, but like physicians they are losing ground to technical, economic, and bureaucratic pressures. Their increased authority in making decisions about their own care is increasingly restricted by economic and administrative constraints that limit the range of choices available in a given hospital or clinic.

TYPES OF PHYSICIAN AUTONOMY

Much of the conversation among physicians today, reflected in professional newsletters, is concerned with malpractice, economic threats of Medicare cuts, and the impending glut of physicians intensified by the growth of non-physician health professionals. The choices faced by newly certified residents are geographically restricted, and expected incomes are declining even as more physicians become dependent on a salary from a large employer rather than solo practice. Each of these developments is perceived as a threat to physician autonomy, and indeed each may be a potent threat. The threats to autonomy are not, however, different from those faced by attorneys, scientists, plumbers, or janitors. The threatened freedom of choice concerns career expectations, social prestige, income, and the political power to control them.

Autonomy in choosing what types of care are available for patients does, in contrast, have ethical implications. Restriction of the range of choices of technologies can affect patients' health. To the degree that physicians are so restricted, patients are worse off, and that matters ethically.

The remainder of this paper will be an elaboration of these hypotheses. First, health technology will be defined and its development traced. Then the way in which technology has changed physician behavior will be addressed. Finally, the consequences of technological advance on reducing the power of individual physicians and patients will be described and illustrated with a few examples.

DEFINITION OF MEDICAL TECHNOLOGY

According to one definition, a medical technology is a drug, device, or medical or surgical procedure used in medical care [20].The circularity and imprecision of this definition testify to the difficulty in generalizing about medical technologies. Medical technologies are defined by their use, and the edges of medical care are impossible to define carefully. Gray areas abound, and agreement on traditional terms is crumbling under the mounting pressures of scientific discovery and technical development. A drug, for example, used to mean a relatively simple and pure chemical used to treat diseases or symptoms. The boundary between drug and biologic has in recent years become quite fuzzy, with the development of neuropeptides, releasing factors, and nucleic acid derivatives. The expansiveness of what counts as medical technology is also contingent on cultural norms. Some would include "quack" remedies, sanitation, diet and exercise programs, experimental therapies, and other practices (e.g., chiropractics, rolfing, meditation, stress reduction, biofeedback), while others would limit consideration to those technologies that are in the mainstream of physicians' practice (itself variable from region to region and individual to individual). Having acknowledged the limitations of defining medical technology, analysis must nonetheless attempt to ignore the difficulties in order to seek illuminating insights through examination of existing technologies in various stages of development.

For the purposes of this paper, medical technology will primarily refer to "mainstream" technologies that are explicitly mentioned. Most of the examples derive from molecular genetics (e.g., genetic screening, prenatal testing, and gene therapy) or are used in diagnosis or treatment of neurological illness. This is primarily because the author is most familiar with these areas. Genetics and neuroscience have the added virtues of both representing rapid technical change and promising to revolutionize larger sectors of medical care over the next few decades.

EVOLUTION OF MEDICAL TECHNOLOGIES

The advance of technology stems increasingly from biomedical research and technological advance in non-medical fields. Discoveries about biological mechanisms often breed new inventions (e.g., prenatal tests for genetic diseases). Technologies developed for non-medical purposes often find novel applications in medicine (e.g., nuclear magnetic resonance for body imaging and fiberoptics for endoscopy). The sources of a particular technology cannot usually be traced to a single invention or discovery. Rather, many disparate ideas, activities, and people conspire to generate a useful technique and the conspiracy can be detected only in retrospect.

Each technological innovation is unique in its genesis, development, adoption, and obsolescence. Yet there are some common elements in the histories of many medical devices, drugs, and techniques. A conceptual model of developmental stages describes the evolution of a "typical" medical technology. Few technologies pass through all stages in the usual order, and some wax and wane in popularity. The relation between user and manufacturer can dramatically alter the direction and rapidity of technical development [1], but a relatively simple model can highlight most of the usual landmarks.

Conception

The conception of a new technology usually arises from one or a few individuals with a good idea. This exalted aspect of technological evolution frequently involves applying knowledge from basic biomedical research or noticing the possibility of adapting an existing technology to a new use. This first step is often taken by those engaged in basic research, frequently in consultation with those engaged in clinical practice.

Experimentation and Introduction

When a new idea seems likely to be useful, a few researchers or firms identify the opportunity and begin to work seriously on bringing the idea to fruition. In somatic cell gene therapy, for example, the relevant animal experiments on gene transfer have been performed in hundreds of laboratories throughout the world, but fewer than a half dozen laboratories are working seriously towards the first reputable human trials. This is typical of the evolution of most technologies. Research on safety and efficacy precedes but overlaps with human trials. Technologies imported from other countries, however, may be directly applied to people. Introduction of radial keratotomy in the United States began in a few clinical centers, as did *in vitro* fertilization with embryo transfer. The more humans involved, the closer the technology is to standard clinical testing.

Clinical Testing

Early experimentation in humans overlaps with formal testing of effectiveness and safety. During clinical testing, the regulatory process becomes engaged, with drugs and devices most often falling under the jurisdiction of the Food and Drug Administration (FDA) in the United States, and comparable agencies abroad. A few techniques escape direct scrutiny by such agencies, and there is some room for discretion by the regulatory agency. (Bone marrow transplantation and *in vitro* fertilization are examples of techniques that generally have not been subject to FDA review.) Review mechanisms for protecting human subjects during clinical testing have been institutionalized mainly in the last decade, following widely publicized violations of medical ethics.

Acceptance or Rejection

After initial successes or failures, a technology is usually either widely accepted or its use becomes progressively less common. The direction of change depends on the experience at the pioneering centers. The minority of technologies that enjoy early success are likely to be accepted, while most languish for lack of encouraging results. Pioneers must balance tenacity against risk. The costs of persistence for academic researchers are usually loss of other opportunies and reduced probability of promotion if the technology is a failure (tenure being an antiquated concept at most medical schools). Financial incentives are most important for proprietary developers. One important set of risks centers on patients. Use of one technology may preclude use of another, especially during clinical trials. Investigators often must balance the need to generate definite information about efficacy and safety against the interests of the patients in the clinical trial group with worse outcome.

The process of review sketched above is, however, the path taken only by technological *goods*. Most technologies – here referring to novel applications of knowledge or existing devices – escape this rational process and become part of medical practice following a few reports of success using a particular procedure. This is true for surgical procedures, novel uses of existing drugs, new uses of diagnostic tests, and new methods of delivering care. The *first* use of a technology is usually much more carefully inspected than subsequent uses. Yet the bulk of innovation in medicine consists of refinements in use of existing technologies, rather than creation of new products. Systematic assessment of new uses for existing technology is a new interest, and the legislative framework and regulatory network for such analysis is not as well developed as that for initial use.

Availability can overwhelm prudence. Genetic screening of large populations, for example, was implemented in many cases before its adverse effect on privacy of individuals was widely understood [27]. The availability of the tests seemed to make their use imperative, and resulted in the propagation of misinformation about sickle cell trait and the unnecessary stigmatization of many people who would have been better off without the new knowledge about their genetic constitution.[4] Acceptance of a new technology involves many factors. The expected risks and benefits are paramount, but social and political factors are inevitably involved as well.

History indicates that some promising technologies are likely to be accepted before they are carefully studied. Coronary artery bypass surgery was widely practiced years before its efficacy was established [2, 11]. The recent difficulties in restraining the widespread use of chorionic villus biopsy[5] and radial keratotomy[6] may also prove good examples of rapid dissemination of new procedures.

The transition from research to clinical trial is contentious for many technologies. Several mechanisms have been used to try to reduce both uncertainty and controversy. Consensus development conferences have been used by the National Institutes of Health [18], and an increasing number of agencies and private organizations house medical technology assessment groups [24]. Yet there is no stable and coherent methodology for assessing technologies that has solved the difficult political and analytical problems. The methods used to generate consensus are themselves often the focus of controversy. The need for technology assessment is widely recognized by government and private payers, but the proper mechanism for performing it is not yet settled [4, 24].

The transition from clinical testing to acceptance is not discrete, and many factors other than consensual data on safety and efficacy can influence patterns of use. Professional zeal, the attractiveness of offering patients a new treatment, and the economic environment can all induce rapid acceptance of new technologies, particularly those used for common diseases that are otherwise incurable.

Obsolescence

Disappearance is the fate of most technologies; it is true with a vengeance in medical technology. The demise of technology can be sudden and dramatic (e.g., the withdrawal of thalidomide from the market), or the technology can fall out of use over the relatively short span of a few years (use of aspirin in childhood fevers, use of pneumoencephalography). Digitalis therapy and appendectomy are conspicuous by their longevity, while most daily medical routines differ markedly from those followed a decade ago. Radioactive tracer studies and the iron lung have by and large been supplanted by newer, more effective, safer, or less costly methods. Yet it is clear from observing clinical practice today that the rate of creation of medical technologies has far exceeded the rate of obsolescence.

FACTORS AFFECTING THE RATE OF TECHNOLOGICAL EVOLUTION

Many factors influence how rapidly a technology is accepted into practice. These include scientific, technical, economic, and political influences.

The *amount of capital* needed to develop a technology increasingly determines which ones finally emerge into practice. Drugs are costly to develop, not only because of the needed research, but because only a small proportion of agents are clinically useful and the costs (both economic and logistical) of demonstrating safety and efficacy are quite high. The more capital needed to develop a technology, the larger the organization required to nurture it. Yet those technologies that are expensive to develop are often also quite lucrative. Inspection of drug company earnings, for example, shows that many companies derive the lion's share of profit from one or a few products.

The *amount of specialized knowledge* needed to use and interpret the technology also influences its acceptability. Electronically amplified stethoscopes are likely to be more rapidly assimilated into routine practice than tools for molecular genetic analysis. This is not because such a stethoscope is technologically simpler than other new devices, but because it closely resembles the familiar.

The *regulatory environment* also influences which technologies are developed rapidly. *In vitro* fertilization and embryo transfer centers have proliferated rapidly across the United States in part because they are not subject to regulation that affects other medical technologies. In a sense, such clinics have reaped the benefit of political sensitivity to the abortion issue. Federal agencies have been loath to embroil themselves in the abortion debate, and have thus left the field open for development free of Federal regulation.

The *public attention* focused on a new technology also affects its acceptance, and this can either accelerate its widespread use or inhibit its development. In the case of organ transplantation, it can be argued that the drama and public fascination with the technology has gone far to overcome the cost concerns of payers whose reflex would be to slow its spread into practice. Gene therapy, in contrast, may have been retarded by the adverse publicity attending the unapproved experiments conducted by Martin Cline in 1980, discussed below.

The *number of patients* potentially benefited also determines the rate of acceptance. Technologies of use to many patients are likely to be more rapidly developed. This is in part because a critical political constituency is quickly formed to promote wide access, and also because of the economic incentives inherent in a large potential market. The positive incentives of a large user population can be vitiated, however, if there is no means of paying for products. Malaria, schistosomiasis, and Chaga's disease have been neglected in clinical and basic biomedical research in relation to their enormous worldwide effects on morbidity and mortality — largely because the diseases are prevalent in poor countries, and payment for vaccines or drugs is not ensured.

The *availability of alternative tests or treatments* is also important. Those suffering from incurable or irreversible illnesses often grasp for any alternative, even one that is risky or untested. Hope can outpace rationality, which in part explains the rapidity with which many cancer therapies and arthritis remedies have been disseminated throughout the population. Quack cures tend to cluster around such incurable illnesses, highlighting the market derived from desperation.

Greater *knowledge about a disease* also makes it more "ripe" for technological intervention. The main reason that adenosine deaminase deficiency is more likely to be treated by gene therapy is not only the fatality of ADA deficiency, the absence of need for controlled regulation of the ADA gene, and availability of a cloned gene for insertion, but also because the cause and mechanism of the disease are relatively well understood (and permit educated guesses about efficacy of new treatments).

Those technologies that can easily *fit into specialty practice* are also more likely to be accepted than those that fall across disciplinary lines or into areas with few medical advocates. For example, computed tomographic (CT) scanning, which displaces more primitive and imprecise methods of analyzing the anatomy of brain dysfunction, has been rapidly incorporated into neurology and neurosurgery. Genetic diagnostic testing and counseling have not enjoyed the same rapid growth, in part because the specialized knowledge about genetics does not fall cleanly into pediatrics, obstetrics, or internal medicine, and the afflictions affect only a small fraction of patients seen by physicians in these primary care specialties. (This is not to ignore the important effects of reimbursement practices and other factors.)

INDIRECT CONSEQUENCES OF TECHNOLOGICAL CHANGE

Emergence of a new technology triggers a concatenation of events that eventually alters medical practice. The primary effect is to allow new actions to prevent, diagnose, or treat medical conditions. Thus begins the ripple effect into other parts of medicine. A new diagnostic technology may permit more refined diagnosis of known conditions or may differentiate new diseases or sub-groups of known diseases. A new therapy increases the incentive to make an accurate diagnosis (so that patients can be sorted into categories likely or unlikely to benefit from the technology). It may also stimulate application to disorders other than those for which it was tested initially. Beta-blockers, for example, are now used not only for hypertension but also for certain cardiac arrhythmias, prevention of migraine headaches, and angina pectoris. The availability of beta-blocking agents has permitted investigation of many new uses, and the list of useful applications continues to grow.

The existence of a new drug, device, or method also incrementally increases the amount of knowledge needed to use it. Proper use presupposes familiarity with indications for use, contraindications, side effects, and usually also underlying physiology and mechanism. The proliferation of technologies has been an important factor (but by no means the only one) responsible for the increased specialization of physicians and biomedical scientists. No one physician can know about the latest in cardiac drugs, neuroradiological imaging, and psychotherapy. Increased specialization is reflected not only in needing more physicians to cover the wider array of knowledge used in practice, but also shifting allegiance among physician groups – from a relatively homogeneous group of clinical practitioners whose interests were largely congruent and well served by national medical organizations to a network of specialty societies only loosely linked by general medical societies. The relative decline of membership in the AMA has been attended by rapid growth of specialty boards, and even establishment of new boards (such as that for medical genetics).

Many new technologies, particularly those that include complicated hardware or require highly specialized knowledge to use or interpret, also create a need to involve more specialized non-physician ancillary personnel. CT scanners, automated laboratory equipment, radiation treatment facilities, and nuclear medicine departments each come with a full complement of specialized physicians, technologists, nurses, clerks, record-keepers, and maintenance contractors.

Increased medical knowledge, specialization, and involvement of new personnel each in turn make the organization of care more complex. Breakdown in one component can inhibit or paralyze smooth operation of a service. The various diagnostic and treatment services used by a patient must be coordinated, and the administrators of hospitals or clinics thus gain power. Those responsible for ancillary services also become more important. When use of a technology has become standard, sudden lack of availability causes chaos. When the SMA-6 machine (an automated laboratory machine that rapidly and inexpensively performs several blood tests widely used in routine care) is on the blink, functions throughout the hospital are affected.

The increased complexity of services and numbers of people involved complicate the management of medical care institutions. One factor is particularly discomfiting: responsibility for deficiencies (or successes) is diffused as more departments are involved in a patient's care, more technologies subject to breakdown are used in assessment and treatment, and more information subject to misinterpretation is created. This is rarely noted when there are no problems in patient management, but the crucial mistakes are difficult to identify when problems arise. Diffusion of responsibility in turn permits elusion of accountability.

The complexity of new technologies and their perceived risks have led to increased regulation. This began in earnest in the early 1960s with amendments to the Food, Drug, and Cosmetic Act, and has since become pervasive in clinical medicine. More recently, regulations have begun to intrude into research as well. Human subjects participating in research are now protected by regulation. New technologies arising from biomedical research in the United States have undergone particularly acute scrutiny [25].

Human gene therapy is a pertinent example. Human gene therapy is the treatment of a human genetic disease using an introduced gene (involving recombinant genetic material). Three historical episodes of work on human gene therapy have been documented. The first was a set of experiments conducted by Stanfield Rogers and his colleagues in the early 1970s. He treated three siblings suffering from argininemia with the Shope papilloma virus, hoping to cure the disease by providing an enzyme activity known to be present in cells infected with the virus. Rogers did not clear the experiments with local Institutional Review Boards (IRBs), because such boards did not exist. There was no government oversight in either the United States or Germany (the two countries where the investigators and patients resided). Rogers apparently did consult several colleagues before undertaking the experiments, but there was no mechanism for formal review [7]. Looking back, the review seems quite casual, but it was the standard of the time.

The experiments of Martin Cline took place in 1980. He treated two young patients – one in Israel and one in Italy – suffering from the blood disease beta-thalassemia. He conducted the experiments just before his local IRB at the University of California at Los Angeles Medical Center denied him permission. There is some evidence that he did so knowing that his protocol would be rejected by the committee [OTA workshop on Human Gene Therapy, September 27, 1986]. Furthermore, he deviated from the protocol that had been accepted by the hospital in Israel (there apparently was no mechanism for formal agreement at the Italian hospital). Dr. Cline thus performed experiments similar to those done by Dr. Rogers, but at a time when standards of conduct were substantially different. The National Institutes of Health (NIH) punished Dr. Cline for his violations by requiring special review of grant applications and termination of his two grants; the sanctions expired in May 1984.

Without entering into the details of the case, it is useful to inspect the review mechanisms for Dr. Cline's work. He needed clearance from one body: his local Institutional Review Board; the Recombinant DNA Advisory Committee (RAC) at NIH had determined it had no jurisdiction because Cline had agreed not to use recombinant DNA, but rather to use unannealed DNA fragments. This also exempted review by his local Institutional Biosafety Committee (IBC), whose responsibility is to monitor compliance with recombinant DNA laboratory safety regulations.

The situation has become much more complex in the last decade. The regulatory and review schema for human gene therapy is shown in Figure 1. Experiments cannot legally be conducted unless all review groups have approved the protocol. The review groups include the local IRB and IBC and a Human Gene Therapy Working Group of the RAC whose recommendations will be reviewed first by RAC and then by the director of NIH before approval. The Working Group will publish the protocol in the *Federal Register* for public comment. In addition, the FDA has exerted its jurisdiction, and is encouraging all investigators to obtain Investigational New Drug (IND) registration before commencing human experiments. Whether any investigator not using federal funds would be *legally* obligated to get more than IRB approval is not clear, but nationally prominent investigators – even those supported by private firms or non-government medical institutes – are unlikely to risk their professional reputations by failure to obtain clearance from both FDA and NIH. There are thus at least 6 gates that must all be opened before an experiment can commence in good faith. Of course, another Martin Cline could perform experiments in violation of the review schema, but would probably be much more harshly treated by NIH, the press, and possibly the courts.

Human gene therapy thus exemplifies a technology for which there has been a high degree of regulation and oversight even before the first prudent human trials have begun. It is extreme in this respect. Several other factors distinguish it from more typical technologies. Gene theraphy is being developed at a time of heavy malpractice litigation and uncertainty about liability for adverse results. Caution is therefore greater than has been the case for other innovations. Gene therapy is a genetic technology, and thus subject to intense public discussion. Its moral origins are suspect; genetic technologies are particularly controversial in Europe, where political action has recently focused on inhibiting the development of genetic screening and biotechnology. The Greens in the Federal Republic of Germany and several other political groups have made explicit statements that biotechnology is a direct threat to personal liberty. It is indeed ironic that fear of eugenics has reached such a level that those with genetic disease may be barred from access to potentially useful technologies. (Anti-eugenic movements are thus fostering coercive social programs that discriminate on the basis of human genetics – perhaps yet another new form of eugenics.) Gene therapy is also technically complex, fits poorly into medical specialty categories (except, perhaps, pediatrics), and will probably be useful only to a small number of patients (hence attracting little commercial interest).

Public and government scrutiny of new technologies is not uniformly applied. Introduction of new drugs now passes along well-established paths through the FDA, but a new surgical technique or a new application of an existing medical technology faces no such barriers. New methods of oversight of emerging medical technologies are fraught with controversy and political positioning [24]. The recent public flaps over interspecies organ transplantation, methods of facilitating liver and heart donation, and artificial heart implantation are conspicuous examples of how policy is not purely rational. The National Task Force on

Shope virus experiments (1970–73)

Informal solicitation of opinions
Trial in patients

Cline experiments (1980)

Application for NIH grant
Compliance with recombinant DNA guidelines (elimination of recombinant DNA from protocol)
(Dis-)approval by Institutional Review Board at UCLA (pending when experiments performed)
Approval by hospital review board in Israel
Trial in patients

Anticipated review process (1986–1987)

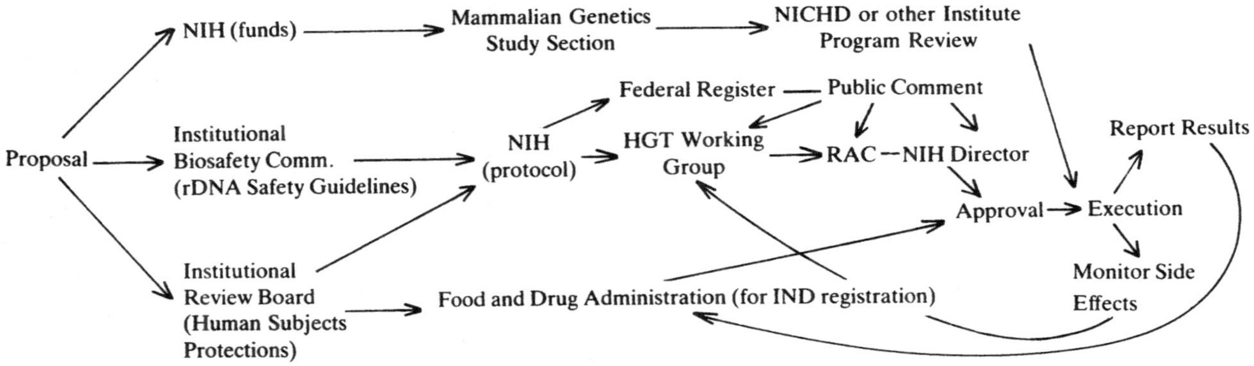

Source: Adapted from [21] with further information from the Office of Technology Assessement, U.S. Congress

Fig. 1. INCREASING COMPLEXITY OF REGULATION AND OVERSIGHT: HUMAN GENE THERAPY

Organ Transplantation and the studies that informed its members, however, emphasize that the rational component *is* important and that dispassionate bioethical analysis can influence outcome.

Need for convenient access to multiple technologies has tended to encourage restriction of new technologies, particularly capital-intensive ones, to hospitals and large clinics. Solo practitioners and small group practices could not finance the costs and could not generate sufficient volume of use to make the investments worthwhile. This has changed recently, but the emergence of surgicenters and CT and MRI scanning services outside hospitals has occurred in response to distortions in how care is paid for, not because of diminished need for convenient access. The pricing and management policies of hospitals have left an opening, and payer policies that encourage outpatient care have fostered community access to care, including major diagnostic and therapeutic technologies.

THE CHANGING ECONOMIC ENVIRONMENT

Medical practices are influenced by the economic environment. Economic factors are clearly not the only factors, however, or even the most important. If they were, then the regional differences in technology use would not be so wide, because financing mechanisms do not vary nearly as much across the United States as do use rates for most technologies.

Economic factors encourage or discourage particular investments with long-term consequences. Prediction based on simple economic analysis is highly subject to error in general, and particularly in health care. For example, hospital stays have shortened under the new prospective payment system for Medicare as expected, but the number of admissions has not increased to compensate (at least in preliminary data) [23]. Retrospective analysis can explain the continued decline in admissions (uncertain profitability of any given admission, displacement of procedures from inpatient to outpatient), but the inaccuracy of prediction highlights the complexity of the health care system and our ignorance of its fundamental dynamics. Some of the factors most important to policy are least subject to direct analysis. The impact of prospective payment on access to care, quality of care, and the health of patients, for example, is even less clear than its effect on aggregate statistics of hospital use.

Economic factors matter nonetheless. Most hospitals have cut staff and rapidly increased their management information systems, and these actions *are* more directly attributable to the new payment methods.

Aside from the debate about financing mechanisms, or rather in parallel to it, is another debate about the shifting economic allegiances of physicians. Paul Starr noted the recent rise of "corporate medicine" in his history of American medicine [31], and Arnold Relman has warned about the potential conflicts between traditional ethical values and emerging incentives arising from new ways to take advantage of economic opportunities [28, 29].

The new ways of organizing health care are not *necessary* sequelae of technological development, but in the current American economy technological advance is fueling entrepreneurial adventurism. This may work to the long-term benefit or detriment of patients, but it certainly thrusts physicians much more prominently into ethical gray zones. Ownership of hospitals, surgicenters, or clinics makes physicians managers and financiers as well as medical practitioners. Physician participation in ownership and management (as well as delivery) of care in the United States will in fact greatly accelerate the shift in public perceptions of physicians away from the image of Marcus Welby and toward J. Pierpont Morgan.

Technology has increased the capital needed to practice medicine. This has increased the amount of capital invested in delivering care, and has raised the financial stakes. Medicine was for a long time a protectorate of mercantilist economic organization, with a guild of physicians shielded from competition by high barriers to entering the profession and distinguished by expertise in an arcane body of knowledge. The public was apparently willing to invest trust in physicians because of a code of ethics that placed patient benefit above other considerations.

Third party payment is one method of avoiding some ethical conflicts. If the patient does not *directly* pay for a service, then there is a weaker conflict between patients' health and economic well-being on one hand, and between the interests of patients and physicians on the other. Insurance and government programs cushion physicians from direct conflict between their medical and business roles. Pressure to restrain cost growth is now transmitted more directly from payers to physicians to patients, compressing this protective cushion.

Other countries have responded to the potential conflicts between the economic interests of physicians and patients by placing physicians on salaries, or by restricting access to certain technologies. In the United

States, however, there is a labyrinthine tangle of financing mechanisms that blends direct public subsidy, insurance financing, indirect tax subsidy, and direct patient payment. The way one pays for one's care depends on employment status, income, age, and type of disease. Some programs are specific for disease category (e.g., mental disorders and end-stage renal diseases have their own programs), age (Medicare), employability (disability under Medicare), financial resources (indigency for Medicaid), and job (large employers tend to have better insurance programs, subsidized by tax policy). These factors all influence care in other countries as well, but not in so direct and obvious a manner.

The panoply of financing methods reflects a lack of consensus about whether health care should be public or private, whether it is an entitlement or a commodity. Moreover, the prevalence of third party payment via fee for service, with little constraint on use, created an environment highly conducive to development of new health technologies. In a fee-for-service third party payment system the resources devoted to health care are not seen as fixed, and so the resources devoted to coronary artery bypass and liver transplantation do not necessarily detract from those used for prevention of polio or long-term care of patients with dementia.

The rise in physician incomes derived from lack of cost restraint. As more dollars were flowing into medical technology, physicians' salaries also rose faster than average wages. Incomes of those most intimately associated with use of technology (surgical specialists and procedure-oriented medical specialists) showed disproportionate gains. If there had been a fixed pie, or if the economic organization of care fostered efficiency, then this disproportionate rise in physician incomes would not have been expected. The unequal rise among physician specialties has further fragmented the interests of physicians along disciplinary lines. The rise of average physician income has also augmented public suspicions of physician avarice.

The power of third party payers also gives rise, however, to a new countervailing force: the discretion of the payer. If a service is not reimbursed, then it is unlikely to be used for long. This control on costs has been invoked with increased frequency as costs have escalated. The emphasis on slowing the growth of medical spending is a political phenomenon. Cost containment is a public issue because the expenses are high enough to attract attention; the size of the medical sector is large enough to interfere substantially with other economic sectors, or at least be perceived to do so. Third party payers, including but not restricted to the government, have responded to pressures to reduce costs by scrutinizing what practices they will pay for. Restraining payment presupposes knowledge about what is medically proper, and payers have been reluctant to exhibit such presumption until recently. But the range of choices available to payers is narrowing as concern for cost escalates. Cost restraint means either directing use of technologies or fixing the budgets of providers, including physicians. In practice, payers have begun to do a little of each. Either choice encroaches on the physician's traditional terrain: the ability to charge what the guild set as the fee or the ability to freely control the use of technology.

The discussion of economic factors could be ignored if economic organization did not interdigitate so extensively with daily care. In the United States, there are conflicting visions of the economics of health care. Decisions are to be left to a "value-neutral" market mechanism wherever possible. Manufacture of drugs and medical devices, for example, is regulated more closely than but not organized differently from manufacture of other goods. The service component of medicine, however, is organized and financed in ways that contrast strongly with pure capitalistic competition; and separating service from goods is becoming increasingly difficult, as goods (technologies) have more and more to do with delivering care. Choices about which technologies are developed now have important moral consequences. In a profit-driven system, economic exploitability becomes the primary determinant of which technologies are developed (as opposed to other possible criteria such as aggregate health benefit or general potential for reducing health costs). The only formal mechanism for aligning ethical norms with economic incentives is government.

To briefly recapitulate, a desire for health begot biomedical research begot new technologies begot higher costs begot cost restraint. The government role in overseeing this reproductive nightmare differs from country to country. In the United States, research has been publicly funded, development of technologies has been largely public at the source but private in application, and costs have been borne by a complex hybrid of public and private means. Health care has for a long time remained a residual enclave protected from large centralized economic organization. Health care is increasingly influenced by corporate organizations with sufficient capital to support continued technological growth. The corporation is supplanting the traditional medical guild. Control of economic decisions was long left to the professional discretion of physicians and their organizations, but rapid increases in costs have induced feedback mediated by payers, both government and private insurers (as well as the businesses that pay for a large fraction of health insurance).

Physicians and patients were the first to benefit from the growth of medical technology, but the large economic stakes are attracting those seeking profit, thereby introducing forces beyond the direct control of either individual physicians or patients. In a competitive medical market, the ethical choices may be obscured by economic ideology. It falls to government to write the rules so that economic incentives encourage ethically correct decisions.

WHAT'S A DOCTOR TO DO?

The role of physicians has changed. The primary care physician whose knowledge could span the universe of medicine has not only disappeared as a reality, but faded even as an ideal. Those entering medical careers now have to choose among disparate paths. Physicians are engaged in primary care, clinical specialty research, or basic research, with much less bridging than used to be the case. Physicians in clinical practice are increasingly remote from decisions made about which technologies are developed, and Ph.D. and M.D. scientists are increasingly disconnected from primary care.

Despite these changes, physicians make most choices about what technologies to use. Their autonomy in *medical* choices is generally not questioned, although their economic and career autonomy are clearly threatened at present. Doctors' orders still ultimately determine which tests are run and which prescriptions filled, but physicians (and particularly primary care doctors) have little say in which tests are available at their hospital or which drugs a pharmaceutical manufacturer chooses to market.

Judgments about trends in physician autonomy thus depend on answering the query: autonomy for what? Autonomy in clinical choices is increasingly regarded as properly left to patients, with the physician as the expert consultant. Choices that affect long-term changes in medical practice (which technologies are developed, hospital administration, clinical business practices) are made by owners and economic interests.

Table I
Increasing complexity and heterogeneity of physician roles

Primary care doctor	Specialist
Clinical scientist	Basic scientist
Personal or family counselor	
Educator	Rationer
Patient advocate	Notary: Assessor of patient eligibility for programs
Surrogate decisionmaker	Clinical supervisor
Businessperson	Employee
Member of the profession	
Case manager	
other specialists	
diagnostic therapeutic services	
patient education resources	
support groups	
public programs	
Product of family, culture, and education	

The overall organization of medical education and health care is increasingly in the hands of government and major payers. Yet, paradoxically, physicians have an increased range of choices in daily clinical care because more technologies are available for use and the role of expertise in clinical decisions remains in the hands of physicians by virtue of their protracted and demanding education.

In addition to consulting in medical decisions with their patients, physicians have a number of other roles, as shown in Table 1, and the importance of non-medical physician activity appears to be increasing with the intrusion of other actors (payers and regulators) and the persistence of chronic disease (requiring patients to need a wider array of services beyond hospital care).

Physicians are counselors and educators, accepting the precepts of their role as navigator rather than captain. They often serve as patient advocates, defending the interests of their clients in the medical system. Doctors are increasingly used as notaries, determining eligibility, or as case managers, coordinating services for patients. This results from dealing with patients with increasing disabilities over long periods, in turn due to the success of modern medicine in delaying death.

Physicians are business managers, financiers, and employees of increasingly large and complex medical institutions. Their roles in such enterprises often conflict with the interests of patients, yet there is little clear distinction among the conflicting roles. It is essential that the same ethical codes that physicians claim when dealing with patients also hold when making financial and institutional decisions: patient benefit should be paramount.

Judgment of patient benefit is not as simple in aggregate as it is for individual patients, however.[7] The difficulties are obvious, while the solutions are not. Yet a simple rule can at least minimize the hazard of direct economic conflict: Physicians should not be in the position of deriving economic gain from their clinical judgments about patients. This principle is included in the ethical code of the World Medical Association, yet practices in the United States violate it.

The cushion of third party payers was salutary in divorcing physician assessment of clinical benefit from economic harm to patients arising from out-of-pocket costs, even though it probably encouraged much unnecessary treatment. The increasing trend to higher co-insurance and out-of-pocket payment (long the norm for long-term care) has revived the conflict between physicians' economic benefit and patients' economic choices. Physician ownership of health care facilities further intensifies it. Those physicians who garner wealth not only directly from clinical evaluation and treatment of their patients, but also indirectly from ownership or management of health facilities, are unlikely to be regarded as ethically neutral or prone to keeping patient benefit above personal financial gain. The equivalent, in the practice of law, to the principle that physicians should not refer patients to facilities or services that they own or partially own has long been the norm among attorneys, and yet it is common for physicians to own shares in diagnosis centers, laboratories, hospitals, or nursing homes to which they refer patients [28, 29]. There is little that can be imagined to cause more permanent damage to the credibility of the profession than this widely accepted behavior. This does not imply that no physician can own shares in any health facility, but rather that a physician should not refer patients to those facilities that he or she is financially tied to, and physicians should shed their clinical responsibilities when they take on the mantle of administration or management as their primary duties (although it would generally be acceptable to maintain clinical interests at institutions they do not manage). These suggested changes are conceptually simple but politically difficult.

The bedrock of medical ethics is respect for the patient's benefit above other considerations. Patient benefit is, however, tacitly restricted to *medical* benefit (or, more precisely, intended medical benefits, as fee-for-*successful*-service is not a payment system in use today). The patient's *financial* benefit rarely enters into the calculations, and that has continued to be a dark corner of medical ethics, illuminated only fitfully by the recognition that patients' own judgments of their best interests – which could include financial considerations – have more ultimate moral force than the judgments of physicians.

When physicians were relatively poorer and the technology they controlled was less costly, the tradeoff between financial and medical benefits was less obvious. Now, costly technologies have brought the conflict to the fore. This has been further intensified by physicians' simultaneous service as owners, managers, financiers, and gate-keepers of medical technologies. Jonsen has noted the central conflict between altruism and self-interest among physicians [14]; technology intensifies this by providing new sources of conflict.

The clash between physicians' fiscal and medical duties merits much greater attention in the bioethical literature. The ethical codes of the American Medical Association, the World Medical Association, and the World Health Organization do not explicitly recognize the potential conflicts between economic and fiscal priorities, or between physician judgment of patient medical benefit and patient assessment of overall benefit. When medical costs are so high, it is much less clear whether a patient is best served by seeing a doctor one more time, having another diagnostic test, or going to the movies ten times. Moreover, as more technologies are used, the likelihood of error and ensuing harm increases. If twenty tests are used, the probability of misinterpreting at least one result is greater than for two tests. As medical care is delivered to the same patient for longer periods by more professionals in more clinics, the likelihood of someone's making a mistake somewhere along the line is much greater.[8]

The attempted divorce of medical benefit from overall patient benefit cannot be effected; the marriage is binding. The only person finally capable of making benefit judgments is the patient – an important reason for locating the power of making decisions in the patient.

Physicians are likely to remain the arbiters of daily use of technologies, retaining traditional roles as patient educator and expert. While patients remain the final judges of proper action, the knowledge possessed by physicians will preserve their power as navigators. The range of medical choices will broaden, but control over career and long-term technological choices will increasingly fall to payers and administrators or evaporate into the oblivion of organizational complexity. Physicians will continue to be crucial in developing new technologies, but different physicians will be involved at different stages. Those in clinical practice will have less to do in deciding about long-term research and development investments. Investigators will be less informed about how their work relates to patient care. Continued technical advance, particularly in an ideological environment that favors "competition," will further exacerbate the tension between the "professional model [that] advocates providing the best possible care at any price, [and the] competitive model [that] advocates providing good care at the best price" [16].

WHAT DOES IT MATTER?

What do the anticipated direct and indirect effects of technological change portend? To whom physicians will be financially beholden – hospitals, employers or others – is not of itself a question of central importance to public policy or deserving of ethical analysis; it is important because of its consequences for patients. Constraining the choices of physicians also limits patients.

From the view of the patient, the physician is merely an extension of the health care system, a means to promoting health. In this sense, physicians are themselves medical technologies that can be used or not. Protection of physicians' economic and career interests has no intrinsic ethical merit. Moreover, it seems quite likely that physicians will be increasingly regarded as mere "inputs" by those who plan delivery of health services as well. While indispensable, physicians are expensive, and thus obvious targets for replacement by less costly personnel or machines. One might even guess that the touted inhibition of technological advance prognosticated as a consequence of prospective payment will be more than compensated by the drive to reduce labor costs (with physician costs being the most greatly reduced). If true, this would encourage development of those technologies likely to replace physician functions, further intensifying the effects of the doctor glut. It would also encourage increased use of physician extenders (e.g., nurse practitioners, physician assistants, and midwives) in lieu of physicians. (The reduced need for physicians projected if HMO staffing patterns become common is one indication that this trend may already have begun [32].)

Given a choice between conflicts (whether financial, legal, or moral) with physicians or conflicts with payers and institutions, one might guess that most patients would prefer to deal with their doctors. The patient-physician relationship is, at least ideally, one between individuals rather than one that pits individual will against institutional policy or computer printout. More important, however, is the primary intent of the care provider. Physicians do acknowledge an explicit moral duty to maintain patient benefit – however defined – as their primary responsibility.[9] Corporations (whether for-profit or non-profit) do not necessarily share this same orientation. The accepted duty of corporations is to maximize their profits on behalf of stockholders, not to enhance the benefit of users. Quality care is a marketing strategy rather than an end in itself.

The primacy of patient benefit in physician codes of ethics makes doctors "vulnerable to enlightened pressure. But if it is to be fully successful such pressure should come from within [the profession] and should be well informed. Ignorant outside pressure will always be angrily resisted" ([13], p. 120). Whether the intention of physicians is truly patient benefit, whether the operative definition of patient benefit is the right one, and whether treating health care as a commodity will lead to decrements in quality are certainly points open to debate. It is nonetheless clear that the values that come into play in highly technological medicine need to be clarified.

The next generation of physicians, whose career expectations will probably not include the presupposition of complete domination of medical care, may well behave more in accord with ethical precepts that emphasize patient autonomy. That generation's subordination to corporate organization may, however, slowly push them away from foremost regard for patient benefit. Whether hospitals, clinics, insurance companies, health corporations, and government administrators will choose to place patient benefit above other interests remains to be seen; ensuring that they do will depend on debate that continually focuses on the ethical judgments implicit in health policy. The ethical values in medicine will increasingly depend on how well government directs the economic organization of health care to concur with ethical values. Medical technology will continue to expand the range of choices available to patients and physicians, to complicate the organization of health care, to make the system of health care less controllable, and to force confrontation with economic and ethical dilemmas.

Office of Technology Assessment, U.S. Congress
and Georgetown University
Washington, DC

NOTES

[1] The views in this paper are those of the author alone, and do not necessarily reflect those of the Office of Technology Assessment or members of the Technology Assessment Board.

[2] This parenthetical point is an elaboration of observations by Carol Winograd, M.D., at the California Task Force on Alzheimer's Disease meeting, University of California at San Francisco Medical Center, January 16, 1986.

[3] Many factors contribute to this trend: lower salaries in research versus clinical practice, continual need to seek new funding for research, the push for primary care in the 1970s, poor and worsening prospects for the health of academic medical centers, the touted viciousness of academic politics, and increasing indebtedness of medical school graduates.

[4] The public perception of sickle cell trait provides several good illustrations. Those heterozygous for sickle cell trait were excluded for several years from candidacy at the Air Force Academy because they were deemed likely to suffer sickling crises. The Black Panthers in Chicago threatened violent opposition to testing, and distributed leaflets that claimed sickle cell trait was as unhealthy as sickle cell disease, and that Blacks were being barred from this information in an attempt to politically minimize the seriousness of the problem (James Bowman, M.D., personal communication, University of Chicago).

[5] Chorionic villus biopsy is a new method for obtaining tissue earlier in pregnancy than is possible using amniocentesis. It was first used in China and Scandinavia, and was also introduced to the United States a few years ago. The national study of chorionic villus sampling in the United States will not be randomized, unlike the one being conducted in Canada. Some ascribe this to pressures to use the techique for those deemed eligible.

[6] Radial keratotomy is a simple surgical technique for reshaping the cornea to improve nearsightedness. It was first developed in the Soviet Union, and was introduced several years ago in the United States. Attempts to restrict radial keratotomy to centers conducting careful clinical trials were blocked by antitrust action against the American Academy of Ophthalmology. The antitrust action was never decided by the courts. The case was settled out of court, with the Academy agreeing to change its public position.

[7] It is not clear whether to use a strict utilitarian calculus that embodies certain "side constraints" (a la Nozick [19]) or a Rawlsian approach (e.g., that taken by Norman Daniels [5, 26]). Even if there were consensus on the criteria for fair distribution of health care, deciding what measurements to emphasize would still encourage confusion and controversy. Should one pay attention to quality, access, preservation of options, or economic efficiency?

[8] The prevalence of malpractice litigation against physicians has shown two surges in recent decades. This is often attributed to the increasing number of attorneys in the

United States and to the increasingly "economic" rather than "professional" public image of doctors. In addition, public knowledge about medicine is much greater than in previous decades: the current adult population in the United States is much better educated and has grown up in an environment of consumer advocacy and challenge to authority. The relative contributions of these factors and others cannot be resolved, but another important factor is often overlooked: the rise of technology, with its consequent complexity and likelihood of error.

[9] Some would divide this into a primary duty to avoid harm and a secondary duty to promote health, but the distinction is often difficult to maintain when technology is involved, and is glossed over here.

BIBLIOGRAPHY

1. Blume, S. S.: 1985, 'The Significance of Technological Change in Medicine: An Introduction', *Research Policy* **14** 173–177.
2. Braunwald, E.: 1977, 'Coronary-Artery Surgery at the Crossroads', *New England Journal of Medicine* **297**, 661–663.
3. Chassin, M. R., Brook, R. H., Park, R. E., etc. et al.: 1986, 'Variations in the Use of Medical and Surgical Services by the Medicare Population', *New England Journal of Medicine* **314**, 285–290.
4. Committee for Evaluating Medical Technologies in Clinical Use, Institute of Medicine: 1985, *Assessing Medical Technologies*, National Academy Press, Washington, DC.
5. Daniels, N.: 1985, *Just Health Care*, Cambridge University Press, New York. See especially Chapter 6.
6. Faden, R. R., Beauchamp, T. L., in collaboration with King, N. M. P.: 1986, *A History and Theory of Informed Consent*, Oxford University Press, New York. See especially Chapters 3 and 4.
7. Fletcher, J. C.: 1983, 'Moral Problems and Ethical Issues in Prospective Human Gene Therapy', *Virginia Law Review* **69**, 515–546.
8. Friedman, E.: 1986, 'Practice Variations: Where Will the Push to Fall in Line End?' *Medical World News* (January 27), pp. 51–69.
9. Fries, J.: 1980, 'Aging, Natural Death, and the Compression of Morbidity', *New England Journal of Medicine* **303**, 130–135.
10. Haug, M. R., and Lavin, B.: 1981, 'Practitioner or Patient – Who's in Charge?' *Journal of Health and Social Behavior* **22**, 212–229.
11. Hiatt, H.: 1977, 'Lessons from the Coronary Bypass Debate', *New England Journal of Medicine* 297, 1462–1464. See also the adjacent correspondence section, pp. 1464–1470.
12. Himmelstein, D. U., and Woolhandler, S.: 1986, 'Cost without Benefit: Administrative Waste in U.S. Health Care', *New England Journal of Medicine* **314**, 441–445.
13. Horrobin, D. F.: 1978 *Medical Hubris*, Eden Press, Montreal, and Lunesdale House, Hornby, Lancaster, England.
14. Jonsen, A.: 1983, 'Watching the Doctor', *New England Journal of Medicine* **308**, 1531–1535.
15. Katz, J.: 1984, *The Silent World of Doctor and Patient*, Free Press, New York.
16. Light, D. W.: 1983, 'Is Competition Bad?' *New England Journal of Medicine* **309**, 1315–1319.

17. Lowrance. W. W.: 1985, *Modern Science and Human Values*, Oxford University Press, New York.
18. Mullan, F., and Jacoby I.: 1985, 'The Town Meeting for Technology', *Journal of the American Medical Association* **254**, 1968–1972.
19. Nozick, R.: 1974, *Anarchy, State and Utopia*, Blackwell's Oxford. See especially Chapter 3.
20. Office of Technology Assessment, U. S. Congress: September 1982, *Strategies for Medical Technology Assessment*, Government Printing Office, Washington, DC.
21. Office of Technology Assessment, U. S. Congress: December 1984, *Human Gene Therapy*, Government Printing Office, Washington, DC.
22. Office of Technology Assessment, U. S. Congress: June 1985, *Technology and Aging in America*, Government Printing Office, Washington, DC.
23. Office of Technology Assessment, U. S. Congress: October 1985, *Medicare's Prospective Payment System*, Government Printing Office, Washington, DC.
24. Perry, S.: 1986, 'Technology Assessment: Continuing Uncertainty', *New England Journal of Medicine* **314**, 240–244.
25. President's Commission for the Study of Ethical Problems in Medicine and Biomedical and Behavioral Research: June 1982, *Compensating for Research Injuries*, Government Printing Office, Washington, DC. See especially Chapter 2.
26. Rawls, J.: 1971, *A Theory of Justice*, Harvard University Press, Cambridge, MA. See especially Chapter 3.
27. Reilly, P.: 1977, *Genetics, Law and Public Policy*, Harvard University Press, Cambridge, MA. See especially Chapter 2.
28. Relman, A. S.: 1985, 'Cost Control, Doctor's Ethics, and Patient Care', *Issues in Science and Technology* **1**, Winter, 103–111.
29. Relman, A. S.: 1985, 'Dealing With Conflicts of Interest', *New England Journal of Medicine* **313**, 749–751.
30. Schneider, E., and Brody, J.: 1983, 'Aging, Natural Death, and the Compression of Morbidity: Another View', *New England Journal of Medicine* **309**, 854–856.
31. Starr, P.: 1982, *The Social Transformation of American Medicine*, Basic Books, Inc., New York. See especially Chapter 5.
32. Steinwachs, D. M., Weiner, J. P., Shapiro, S. et al.: 1986, 'A Comparison of the Requirements for Primary Care Physicians in HMOs with Projections Made by the GMENAC', *New England Journal of Medicine* **314**, 217–222.
33. Wennberg, J.: 1986, 'Which Rate is Right?' *New England Journal of Medicine* **314**, 310–311.

OTHER REFERENCES CONSULTED

Heilbroner, R. L.: 1985, *The Nature and Logic of Capitalism*, Norton Press, New York.
Ziman, J.: 1978, *Reliable Knowledge: An Exploration of the Grounds for Belief in Science*, Cambridge University Press, New York.

STUART F. SPICKER

MARKETING HEALTH CARE: ETHICAL CHALLENGE TO PHYSICIANS

> The failure to rationalize medical services under public control meant that sooner or later they would be rationalized under private control. Instead of public regulation, there will be corporate planning. Instead of public financing for prepaid plans that might be managed by the subscribers' chosen representatives, there will be corporate financing for private plans controlled by conglomerates whose interests will be determined by the rate of return on investments. That is the future toward which American medicine now seems to be headed.
>
> But a trend is not necessarily fate ([42], p. 449).
>
> PAUL STARR

I. INTRODUCTION

Physicians and the public are becoming uneasy, if not hostile, toward the idea of health care marketing [1, 2, 5, 7, 12, 24, 37]. Yet virtually everyone agrees that certain aspects of managing the practice of medicine and the other health professions belong to the realm of business – that health and medical care share certain similarities with other services and goods that are provided in an open market economy in which expenditures are determined by the product of cost and volume.

The struggle now underway on the deck of the health ship afloat on the already rough sea of the organization and financing of health care suggests that the prudent stay below, remaining "passive" (like complying patients always have) [3], and leave the altercation on the bridge to the medical, managerial, insurance, governmental, and legal forces. But "prudence" often masquerades for cowardice, and it may be less than

Nancy M. P. King, Larry R. Churchill, and Alan W. Cross (eds.)
The Physician as Captain of the Ship: A Critical Reappraisal, 159–176.

virtuous to remain comfortable below; after all, everyone aboard the vessel is a potential patient, and given the odds of the natural lottery, who knows who will be seasick next, perhaps even chronically or terminally ill? If you climb to the upper deck, however, you not only may witness the struggle on the bridge, but also observe a clash of other flotillas on the ocean surface. The very astute may glimpse some underwater currents that surely affect the apparently surface altercation: among the undercurrents are the shareholders of for-profit corporations and hospitals, typically detached from everyone, yet in the end committed to influencing all ships if their earnings are in jeopardy, whether or not physicians are at the helm.

Amid such vexatious uncertainty there is always room for humor. In December, 1985, two physicians, Drs. James V. Maloney and Keith Reemtsma, outlined a strategy that would rapidly assist those waging battle over increasing expenditures for health care, but they admitted it was drastic and "a bit gross," and they really could not prescribe it for a peaceable society. They suggested that President Reagan's major objective in invading Grenada "was to close St. George's Medical School (although Russians, Democrats, and M. Mitterrand attributed to him more sinister motives)" ([22], p. 1713). They calculated, "If the medical school in Grenada had remained closed for 10 years, the Reagan administration would have saved the U.S. taxpayers more than $7.9 billion."[1] Their estimated calculation (given 50 graduates per year, with average time in practice of 35 years, at some estimated dollar figure in the half million range) is perhaps unrealistic; however, they go on to weight this *non-expenditure* (too loosely called "savings") against the Department of Defense's estimate that the cost of this invasion by naval armada was $134.4 million (not including military pay). So they conclude that the outcome was a "favorable benefit-cost ratio of 59:1" ([22], p, 1713).

This scenario, though whimsical, is, oddly, not radical enough, for "spending for health in the United States reached $387 billion in 1984, amounting to 10.6 percent of the gross national product" ([20], p. 1), and if the 1984 rate of expenditures had been allowed to continue, by 1990 the expenditures for health care would have reached $600 billion per year. (Such figures typically fail to stir us until we appreciate the fact that it takes $1,000 per day for 1,000 days [2 years and 9 months] to fill one's coffers with $1 million; $1 million per day for 1,000 days to acquire $1 billion; and $1 billion per day for less than one year and two months to pocket $415 billion, which is above the present 10.6 percent GNP expended for health care in the U.S. These figures speak dramatically for themselves, even if the billions expended on health care are compassionate.)

Any attempt to appraise the traditional role of the physician as captain of the "health armada" (or at least the "medical fleet") requires that one evaluate the present struggle and the public's evaluation of physician conduct, given the recent movement radically to refinance health care in the United States. For in all likelihood, it is the physicians who will be expected to play a major role in the complex social movement to control expenditures for health care in a newly structured competitive setting.

Properly to approach the new health care management model one should consider what a proper marketing of medical or health care could be like, where truly competitive possibilities are retained. A sound marketing approach might, in the end, be a boon, actually assisting in the provision of high quality patient care, in addition to eliminating the present inefficiencies and actual waste of human and material resources. Certainly this possibility is not intrinsically incompatible with the new for-profit scenario. Thus, if physicians ask, "How can we provide what the public wants from their health care system – access, responsible expenditures, quality care, and integrity of professional-patient interactions?" rather than, "How can we be sure we profit significantly from the services we provide?" (not to mention, "How can the stockholders profit so that our shares remain attractive?"), the future might indeed be better than the present.

But the ethical challenges facing physicians (as they take part in various health care plans and prospective payment systems) may well reduce to the ways the principles and practices of sound medical action can be preserved. At the same time, physicians must avoid unscrupulous profiteering and advertising, questionable methods in the competition for patients, and other practices that compromise those clinical values that they claim have sustained the practice of medicine and their care of patients for centuries. Moreover, the managers and insurers [4] have an ethical mandate as well: to leave the practice of medicine to the health professionals and clinical judgment to clinical peer review. At this point, a critical qualification is needed: I am *not* arguing in this paper that a for-profit health care market system is the best of all systems in comparison, for example, with Great Britain's National Health Service or

Canada's National Health Insurance scheme. I simply presume that for-profit institutions will increasingly supplement for-profit solo practitioners and non-profit institutions in the U.S. and that, given this, a reconciliation is possible, in principle, which avoids reducing medical and health care to a matter of mere business management. But before this can be argued convincingly, other concrete concerns must be addressed.

II. THE DILEMMA

The literature since Dr. Arnold S. Relman's seminal article of 1980 on "The New Medical-Industrial Complex" [37] is replete with prudential as well as ethical concerns pertinent to physicians and non-physicians. A number of these worries can be reduced to the following dilemma [18]:

(Premise 1) If physicians attend primarily to the needs of patients, serve as their principal advocates, act on the basis of the traditional moral maxims of medicine, and make all medical decisions irrespective of cost, then excessive costs will result, which will lead to bankruptcy in budgets that direct health care services, leading to a still more unacceptable GNP for health care. And, if physicians do not attend first and foremost to the patients' needs, but rather to the group's or third-party's profits, then they surely fail in their moral and professional conduct and, as a secondary consequence, further damage the interests of those lacking health care insurance.

(Premise 2) Either physicians address their patients' needs as their primary concern or consider their profits and income as well as the solvency of their corporate group as their primary allegiance.

(Conclusion) Either bankruptcy of the financial base for U.S. health care will result, or physicians will act unethically, and the public will lose all confidence in the medical profession.

I intend to show that this dilemma is refutable, since its premises are false. Tacit expressions of this argument are common in the literature critical of the transition from public and non-profit to for-profit health care organizations. These critics claim that a moral dilemma is created when physicians act both as patient advocates and as corporate employees and/or investors in for-profit health care organizations. Critics conclude that physicians "have adopted a new set of moral values to guide health policy and medical practice" ([30], p. 919), as if solo practitioners in the past were not profit seeking but viewed their incomes as ignoble. The judgment is made that when the new corporatization of medicine is complete, after it moves rapidly from public and non-profit to for-profit models [38], non-physicians will believe that "they can assume most of the functions of physicians" [29] and thus displace the *power* of health professionals. Physicians' autonomy will be further eroded, since they will become "subservient to insurance companies or third party payers and to the business tycoons who run worldwide hospital chains" [29].

This "standard view" (as I shall call it) of business in health care is even particularized as to wrongdoings, for it has been alleged that clerks have changed physicians' medically correct diagnoses under the present Diagnosis-Related Groups (DRG) system (which is tied to prospective payment models) in order to enable their hospitals to receive more lucrative payments to increase revenues. To be sure, such individual instances of fraud do not prove that the turn to for-profit health care services is immoral, but they do illustrate the growing concern that is reflected by Dr. Iris F. Norstrand: "Why," she laments, "a once proud and independent profession has permitted this tragic takeover [by non-physicians] to occur is indeed incomprehensible" [29].

The beginning of a response to her question causes one to attend to her notion of an "independent profession." That *independence*, even to the extent that it existed in fee-for-service medical care, is less and less available. One might well argue, moreover, that the very complexities of health care financing have made the independent actions of physicians risky for, and not particularly advantageous to, their patients, either; for independence of the sort Dr. Norstrand has in mind may be the tacit expression of her desire for *greater autonomy*, in contrast with the impotence of patients, government officials and taxpayers. It might be useful to note, however, as David Mechanic does, that "doctors have typically been ingenious in responding to reimbursement incentives and in measuring clinical productivity" ([27], p. 182). One could imagine, then, that the turn to large health care service organizations will simply require greater knowledge of the many ways one can re-tune and fine tune the system as well as reward merit – strategies commonly practiced in other corporate organizations but perhaps appearing incompatible with autonomy. Nonetheless, Dr. Norstrand's concern, that non-physicians are gaining too strong a foothold on the very functions of physicians, deserves to be addressed further.

III. DIVISION OF FUNCTION

The move to for-profit, prepaid, corporate ownership of health care services and material resources brings at least three key groups into conflict. It is essential that they eventually learn to divide the critical responsibilities among themselves: managers, physicians (or hospital staff), and patients will have to act out their mutual interests in a series of contexts in which health care services are provided [41].

A. Managers

The turn to for-profit models of financing health care will lead to an increase of non-physician managers, who must, in the end, become more competent managers, for their concern will not only be to discover ways to maximize the profit margin and maintain solvency of their institutions within the ethical norms of the health care context in which they will work. A consensus among physician participants will eventually emerge on various issues, as policies begin to take shape and become known to the clientele in the community. Managers will have to be skilled in data acquisition and analysis [33] and be prepared to work diligently with physicians to determine what constitutes necessary or marginal treatments, or critical or marginal benefits in terms of acceptable and unacceptable risks to patients, given the need to monitor cost-containment and profit goals. (We have already learned, for example, that sometimes fewer laboratory tests may cost more in the long run [16].)

Further, managers will have to accept the task of informing all subscribers to their plan (as well as potential subscribers) of the nature of its arrangements, policies, coverages, limitations, and exclusions, especially since consumers and employees do not always select their health insurance plans (employers may make the selection for them) and, furthermore, since *third party payers serve as mediators between employer-purchasers and employee-patients*. Most importantly, the managers have the task of weighing incentives such that the tendency to *overtreat* patients (as in the previous retrospective fee-for-service payment system) and the new tendency to *undertreat* patients (as is anticipated for the prospective payment systems) are avoided [34]. And managers will have to think creatively to generate non-monetary rewards and thereby to recognize outstanding clinical practice ([27], p. 182). In short, a very special "corporate culture" will have to be created, because health care services, though in many ways similar to other human services, are unique in a few critical ways, e.g., there are professionally self-imposed obligations to attend to the lack of access to health care services for those medically uninsured. (I shall return to this shortly.)

We should take Dr. Norstrand's warning seriously, and establish mechanisms so that non-physician managers do not overstep their mandate, function, and competency. A few years ago, the economist Alain C. Enthoven cautioned that it "remains to be seen" whether the new investor-owned HMOs "can develop the kind of corporate culture that is compatible with high quality care" ([9], p. 1530). Clearly, the newly-formed HMOs and PPOs are having little difficulty recruiting competent young physicians, but "managerial know-how is in short supply" ([9], p. 1530), and this may too easily permit non-physicians to carry out the functions proper to physicians and hospital staff.

B. Physicians

Given the recent incremental rise in power of private third-party payers (insurance companies, employers, and various health care plans), physicians are likely to feel more threatened with further reduction of their traditional level of power and autonomy – which, of course, was never total [6]. Autonomy can be threatened by physicians' fear of reductions of their future incomes should they contribute excessively to the overutilization of resources ([28], p. 11). (This fear was implicit in the dilemma noted in Section II, i.e., the profit-oriented health insurance agencies may be seen as precipitating a reduction in overall quality of patient care, and interfering with the physician-patient relation in particular.) In short, physicians are now wondering whether their decision making will be "over-managed". What then can they do? Better yet, what *ought* they do, given their societal mandate and the danger that managers may not in time establish a trustworthy corporate culture?

First, physicians have to examine critically their own financial involvements. Is it proper for them to own shares in health care institutions in which they are professionally involved? On the other side, can it be argued that they have an obligation to invest in health care in order to (1) sustain the medical care ethos and (2) make sure these institutions remain attentive to the values of medicine? What constitutes a conflict of interest for the medical profession as it seeks to regain and maintain the public's trust and confidence?

Second, physicians must organize to keep the managers at a proper distance from the physician/patient interaction, and to preserve their clinical judgment by being prepared to defend it within the community in which they practice, and even in the courts. This will require, third, that physicians establish reasonable standards of care and communicate them effectively as local policies in order to avoid charges of deceit or coercion by the subscribers (who, by the way, are not currently ill or in need of care, but will be patients at a later time).

Fourth, future policies with regard to patient care visits will require that a reasonable duration be determined for these visits. In this context, fifth, physicians will have to determine as a group, but *prior* to providing patient services, what, if any, the limits of care will be, and then announce them in advance. This requirement will have to include their commitment to provide emergency care ([1], p. 18). Sixth, as part of this requirement, physicians will have to organize to ensure that the group to which they belong (HMO, PPO, etc.) clarifies its patient care obligations *in advance*. If the public ought to be required to familiarize itself with various advance directives for treatment/non-treatment decisions at a later time in life (as I believe it should), then it might be prudent for physicians to consider establishing their group's policies in advance as well.

Seventh, physicians should lobby to achieve cost-effective health insurance coverage [8] for the indigent at the national, state, and local levels [17], and participate in negotiations to enable the indigent to pay insurance premiums. After all, the for-profit movement necessarily results in the decision of many third parties to refrain or withdraw from supporting those who cannot pay for their care. (Here we are referring to 9–13 percent of the U.S. population, between 21 and 30 million people.) This should be anticipated in a market economy, but in my view we should counter this expectation by arguing that health care services should be distinguished from other "goods" and services; this would require an additional set of arguments rooted in our communal belief in a minimal level of care for everyone.

Most importantly, eighth, physicians must distinguish two senses of 'gatekeeper' – the first is the power of the physician-captain to control an individual patient's auxiliary services and the physician's own costs based on his diagnosis; the second is the managerial assessment of limited resources expressed in management's prior financial plan, where overall expenditures and revenues are calculated to yield reasonable profit and solvency, which must be periodically monitored and reviewed by physicians and medical staff in concert with the managers. In short, physicians should formally reject the role of rationer or gatekeeper of resources at the physician-patient level as much as possible, yet be willing to participate in rationing policy management conferences in which hospital or group *policy* is formed; for it is the latter that must be formulated in committee *prior* to individual patient care models.

Given this dual sense of 'gatekeeping,' I must disagree with David Mechanic, who prognosticates that "changes in financing will shift the physician's role from advocacy for the individual patient to greater responsibility for the allocation of fixed budgets" ([27], p. 183). This need not occur, and dismantling of the DRG classification tied to prospective payment could eliminate this danger almost entirely. One can as easily conjecture that proper arrangements and division of function can enable physicians to withdraw from the role of financial, cost-containing gatekeeper by putting mechanisms into place which require management to be essentially responsible for this function. For example, if every PPO, HMO, or hospital does not purchase all available and highly expensive apparatuses, one need not necessarily budget for them – either by way of purchase or by paying to maintain such expensive technology; this is an old illustration, to be sure, but still useful.

In short, the "culture" of owners and managers of for-profit groups will have to serve as gatekeepers (in response to social referenda perhaps) for expensive and specialized services. It is, in the end, a quite different state of affairs when a hospital or community *policy* indicates in advance that, for example, premature infants of body weight less than 800 grams will not receive long-term and expensive life-saving care, than it is for a physician to be required to bear the burden of this decision in virtual isolation. Patient/community advocacy and cost-effectiveness advocacy can thus begin to operate fairly, especially in our present market economy, where not only scarcity and limited resources prevail, but where rationing of health care services and material resources will occur and which therefore will necessitate tragic choices [11, 39].

The goal of managers, owners, and physicians, taken together, is to "distance" the clinician from the economic calculations of medical practice, retaining this critical function for the managers. This distancing will have the important psychological consequence of diluting the effect of focusing on each patient's cost of care and thus the managers will begin to contrive ways to control the purse further by capitation models. (DRGs will, as I predicted at various meetings and conferences a few years ago, soon be a thing of the past.)

Finally, and most importantly, formal mechanisms must be established to require physicians to sanction those within their profession who consistently fail to meet established standards. Failure to scrutinize the medical profession, and all health care professions for that matter, will worsen the situation, for it will encourage malpractice litigation, which is, in the end, always paid for by the patients [44].

As Robert U. Massey, M.D., has remarked: "As physicians. . .we can seize the opportunity to reshape medicine closer to the principles which have animated the best of the profession for 25 hundred years. If a dozen corporations are going to run things, it is our job to be certain that they manage the enterprise in keeping with the highest principles of scientific and altruistic medicine" ([25], p. 207).

Notwithstanding this commitment, one might still offer a cautious qualification: we must not create a system that relies too heavily on *altruistic* conduct[2] of any group any more than we should rely too heavily on private philanthropy to support care for the medically uninsured. Indeed, the occasion is at hand to develop and implement public and health policies and to pay for the safest standard of care for those who are truly unable to pay out-of-pocket for themselves. Even if some persons are prone to abuse the system and take a "free ride" through it, that is no warrant to refuse to provide taxpayer monies for those who are truly in need.

In sum, physicians are responsible to remain alert to the following: ancillary services might become overpriced by the managers to the detriment of patients; managers might make decisions that put patients at excessive risk by reducing the number of hospital personnel per patient; managers might not only tend to "skim off" the so-called "most desirable" (able to pay or be reimbursed) patients from the community or list of subscribers, and "dump" the so-called "less desirable" (uninsured or unable to pay) patients, but one could argue that they would be consistent in doing so in a market economy. Finally, if the Oath of Hippocrates has anything of meaning left in it (and perhaps it does not, owing to its simplicity and the complexity of our world), surely physicians have some obligation to resist decisions that will compromise the quality of medical education.

All of the above demands on physicians could come to naught if careful *contractual agreements* between employers, managers, subscribers and physicians are not drawn. Fear on the part of those who reject the new "profit ethos" can be a useful emotion; it can well encourage the formulation of very specific and legally binding contractual arrangements [35].

C. Patients

First and foremost, patients have a responsibility to evaluate their life styles. If a person continues to live in ways that favor illness, and if that person is not paying out-of-pocket in some fashion for his or her health care services, then he or she has virtually no moral claim against the rest of us to foot the bill. (Here, of course, we should be reminded that excessive smoking might paradoxically reduce total societal health care expenditures through reductions of social security payments due to premature death of heavy smokers.) But patients ought to insist on mechanisms for voicing complaints; ought to insist on the freedom to "shop around" for health care providers and facilities they prefer, if *they* can in part pay for them; ought to refuse to be permanently locked into a given medical care organization, each with its own limited panel of providers; ought, finally, to refuse to remain uninformed, and thus to require clearer information transfer from interpersonal communication with physicians and paraprofessionals. The grave danger, of course, is that employers will retain the power to make those decisions affecting choice of providers. Employees, therefore, must organize to counter the forces to keep them helpless and irrelevant. Here, labor unions may again become active employee advocates.

IV. A WORD ON COMPETITION

The emphasis today is clearly on reducing federal intervention in financing U.S. health care expenditures [10], and increasing market activities to yield cost reductions and other goals [13]. The process of competition, long familiar to us from classical economics, is clearly central to relatively open markets: U.S. antitrust laws were in fact designed to protect citizens from monopolies and oligopolies. Thus, it would be an error to construe the most important and immediate problem as one of further establishing a competitive market among health care service organizations. The focal problem, as Enthoven has concluded, is that "public policy is still far from the competitive model" ([9], p. 1529) in which employers, for example, would require coinsurance and deductibles of their employees, so that every employee pays something out-of-pocket. Such a suggestion, however, quickly finds its critics. Dr. Relman remarks that "we will not know whether cost sharing endangers the health of patients" ([36], p. 1435).

In short, an imperfectly functioning market is still, for most of us, preferable to no market at all. But in saying this one cannot ignore government responsibility in the present rapid transition to broadly based for-profit health care units. "If employees and government want the competition between the prepaid group practice and its competitors to be fair, it is up to them to set the rules so that health plans either cannot select risks or are systematically compensated for caring for high-risk persons" ([9], p. 1529).

V. THE CONCEPT OF THE COMMONS IN HEALTH CARE SERVICES

The concepts of competition and self interest, which have heretofore been proffered for their positive consequences in a capitalist economy [14], contain a dangerous side as well. Self interest can easily lead to greed if not held in check; competition can generate rapid expansion and overutilization of resources so that in that end we, or others to follow, are bankrupt and destitute.

As many recall from the writings of the ecologists, "the commons" refers simply to a resource to which a population has free and *unmanaged* access. In our context, of course, the resource is not a plot of land but the utilization of medical and health care resources, which are clearly subject to distribution by claimants like ourselves. Overdemands made on the medical and health commons (which is here synonymous with free health care provided by third-party payers, who also hope to reduce or at least control expenditures through competition) yield a gain to some individuals but also a loss to be shared by everyone in the entire population under third-party subscription. We may well be fast approaching a time, then, when it will be argued that many of us are "stealing" from the commons—too many liver transplants; too much expended on very premature infants or on maintaining brain-dead bodies, etc. Since stealing by everyday social standards is wrong, perhaps we ought to discontinue those practices that lead to it. To do so, we must support private ownership and regulate competition, and yet regulate access to health care services and all medical resources.

Here again we cannot rely on each individual to think and behave prudently (morally?) regarding the commons. To avoid what Garrett Hardin has dubbed the "tragedy of the commons," it must be controlled by private ownership – we must, in short, ease into "commonizing" expenditures across co-paying subscribers in the third-party health care plans and groups. This in all likelihood will include federal, state, and local *taxation* of all citizens whose resources can be proportionally directed to the medically uninsured. The problem, of course, is that for the libertarian, taxation is another and even more invidious form of stealing, but this matter cannot be addressed adequately here. The difficult task ahead is to reconcile fairly (1) access to health care services for those in need with (2) the limited ability of the commons to serve all medical needs. This is the well-known and ever-present tension between compassion for the individual and bureaucratic efficiency toward the abstract collective; unique to health care is our recent realization that the resource represented by the medical commons is a limited one.

The difference between pathos and tragedy, you may recall, is given in the parable of the protagonist on whom a statue falls, which accidentally kills him; this is pathos. When the statue that kills him is made in his own image, that's tragedy. In tragedy what ends unfortunately is not only foreknown but self-propelled. Is that to be our fate, or can there be a happier ending?

VI. CONCLUSION

With the rapid movement to increased competition in the health care service sector, antitrust legislation and the agencies that oversee it – the Federal Trade Commission (FTC), as well as the antitrust division of the Department of Justice – will become much more active and visible than in previous years (their efforts have waned under the Reagan administration). Recall that not too long ago the FTC took its place on the health care stage (though some critics are convinced that its investigations were far too secretive) [32].

Antitrust legislation, as is well known, was originally designed "to see to it that competition functions freely throughout our economy as the most efficient and least intrusive regulation of a market behavior" ([32], p. 238). Generally speaking, antitrust laws were supposed to assist in minimizing waste [15], increasing efficiency, providing incentives for innovation, and decentralizing decision making. But health care services are *not identical* to other services and the sale of goods, since the essential valuing of persons is intrinsically involved in ministering to the sick; thus, serious questions remain with regard to the application of current antitrust legislation in the health care industry. The fact that group health insurance models, for example, are rapidly expanding in the U.S. is one way to increase competition, for there is no particular legal limit put on the number of such groups and corporations that may be established, and those destined to fail are unlikely to be protected by federal intervention. But to what degree are individuals' choices being restricted in the process? If physicians are to determine their own reimbursement within health insurance groups, is that a violation of antitrust legislation? Are new alternative regulations required to protect the public from monopolies and oligopolies? (Such questions, though critically in need of attention, cannot be considered here. To treat this matter competently and thoroughly I must defer to those experts in health law who understand the mechanisms of various regulatory agencies.)

The fantasy of unlimited health care services and medical resources – of an inexhaustible commons – is the worst possible image, even if sound management principles are not ignored in future practice. The managers – as well as public health policy makers – must now say "No" [43, 45], while physicians must do all they can rationally to say "Yes" whenever sound clinical judgment demands a "Yes." When dramatic conflict ensues, managers, medical staff, and patients must bring their political clout to bear for the good of the patient, knowing that compromise may be unavoidable and some dreams must remain unfulfilled.

In this, one must question as too simplistic Dr. Relman's conclusion, "The best kind of regulation of the health-care marketplace should therefore come from the informed judgments of physicians working in the interests of their patients" ([37], p. 967). Although physicians' involvement is extremely important, management will have to educate physicians in devising performance incentives for salaried practice. The purpose of the efforts of managers and physicians will be to affect the public's perceptions of physicians, so that it comes to view them as much as possible as "disinterested trustees" of patients. This is perhaps less likely to occur if physicians continue to become shareholders in health-care businesses investing, as some see it, in the misfortunes of others.[3] (But the very opposite may turn out to be true; here I am frankly uncertain.) In sum, new legislation may be required to reflect the rules on the basis of which practicing physicians may be permitted to engage in health-care-related enterprises. So, much needs to be done in order to acquire the evidence that supports or refutes the claims that (1) for-profit medicine and health care will not serve patients' expectations for quality care; (2) clinicians will have less and less discretion in for-profit enterprises; (3) physician incomes will shrink dramatically. On the other side, where is the evidence that physicians' basic motives are or must be altruistic? Could that be simply lightly nostalgic?

In conclusion, I have placed in question the common assumption made by many critics of the movement from public and non-profit to for-profit health care models: that there is an "unavoidable conflict between the canons of business and the canons of medical morality" ([35], p. 196). A profit has always been realized from the misfortunes of illness, disease, and disability, and that is not new, though many find it quite troublesome. Of grave importance is the need to establish legal, contractual, and regulatory mechanisms to control excessive self-interest, which can go unaddressed by an excessive preoccupation with managing health care services in the corporate context. To be sure, the legitimate interests of physicians, managers, patients, and other sectors of society must be protected; a balance of power can be achieved such that each party has some degree of control over the functions of the other [19]. After all, physicians obtain their mandate to practice through their patients. It is licensure, a mechanism for restraint of trade that carries with it a virtually unique social and moral warrant, which enables them to treat patients. But physicians are not quite like the factory workers of years past whom the corporate bosses could remove and replace with the indigent hopefully waiting for work, while labor established trade unions as one counter strategy [23]. In our shipboard scenario the patients below may well mutiny and support their own trade unions, throwing their weight against owners, managers, and physicians alike. But the image of the worker-victim captured by Charlie Chaplin in the memorable "Modern Times" is simply inappropriate for licensed health care professionals, given their knowledge and authority. Physicians and other stockholders, acting through the managers, will not, we hope, be able simply to throw overboard the disenchanted and rebellious physicians on the bridge and then simply pirate additional physicians and health professionals to serve on the health ship; or, if they can, it will be clear testimony to failure of the physicians on the bridge to stand by each other when navigating on the stormy economic sea. In 1803, Thomas Percival was not only insightful for his day but indeed prescient when in his *Medical Ethics* he referred to the relationship between medical practitioners as a relationship between the "brethren" ([31], p. 63). Surely you get my drift. . . .

School of Medicine
University of Connecticut Health Center
Farmington, Connecticut

ACKNOWLEDGMENT

I am grateful to my friends and colleagues, Professors H. T. Engelhardt, Jr., Thomas Halper, Nancy King and John Glasgow, for their very helpful criticisms and suggestions that were incorporated into the many revisions of this manuscript.

NOTES

[1] In their letter to the Editor of the *New England Journal of Medicine*, D. W. Light and M. Widman calculate the "savings" of closing St. George Medical School at $32.6 billion, since in their view Drs. Maloney and Reemtsma failed to include in their calculation "the services not ordered during the entire career of each physician not trained" ([21], p. 1548).

[2] Smith's words speak eloquently to this point: "As every individual . . . endeavours as much as he can both to employ his capital in the support of domestic industry, and so to direct that industry that its produce may be of the greatest value; every individual necessarily labours to render the annual revenue of the society as great as he can. He generally, indeed, neither intends to promote the public interest, nor knows how much he is promoting it. . . . [H]e intends only his own gain, and he is in this, as in many other cases, led by an invisible hand to promote an end which was no part of his intention. By pursuing his own interest he frequently promotes that of the society more effectually than when he really intends to promote it. I have never known much good done by those who affected to trade for the public good. It is an affectation, indeed, not very common among merchants, and very few words need be employed in dissuading them from it" ([40], p. 423).

[3] Dr. Relman's views are rather extreme, if not a bit simplistic: In 1980, he maintained that "if physicians are to represent their patients' interests in the new medical marketplace, they should have no economic conflict of interest and therefore no pecuniary association with the medical industrial complex" ([37], p. 967). But physicians will always have either a financial incentive to *overtreat* or to *undertreat*; disinterest is impossible. This is not to say, of course, that financial incentives are the only ones (or even the strongest ones).

BIBLIOGRAPHY

1. Annas, G. J.: 1985, 'Adam Smith in the Emergency Room', *Hastings Center Report* **15** (4), 16–18.

2. Arthur, R. J.: 1985, 'Public Perceptions of Medicine: Letter', *The New England Journal of Medicine* **12** (17) (April 25), 1130.
3. Blendon, R. J. and Altman, D. E.: 1984, 'Public Attitudes About Health-Care Costs: A Lesson in National Schizophrenia', *The New England Journal of Medicine* **311** (9) (August 30), 613–616.
4. Brailey, A. G.: 1980, 'The Promotion of Health Through Health Insurance', *The New England Journal of Medicine* **302** (1) (January 3), 51–52.
5. Califano, J. A.: 1986, *America's Health Care Revolution: Who Lives? Who Dies? Who Pays?*, Random House, New York.
6. Davidson, C.: 1984, 'Are We Physicians Helpless?', *The New England Journal of Medicine* **310** (17) (April 26), 1116–1118.
7. Dobson, R.: 1986, 'Conflicts of Interest and the Physician Entrepreneur', *The New England Journal of Medicine* **313** (4) (January 23), 250.
8. Doubilet, P., Weinstein, M. C. and McNeil, B. J.: 1986, 'Use and Misuse of the Term "Cost Effective" in Medicine', *The New England Journal of Medicine* **314** (4) (January 22), 253–255.
9. Enthoven, A. C.: 1984, 'The Rand Experiment and Economical Health Care', *The New England Journal of Medicine* **310** (23) (June 7), 1528–1530.
10. Fein, R.: 1985, 'Choosing the Arbiter: The Market or the Government', *The New England Journal of Medicine* **313** (2) (June 11), 113–115.
11. Fuchs, V. R.: 1984, 'The "Rationing" of Medical Care', *The New England Journal of Medicine* **311** (24) (December 13), 1572–1573.
12. Fuchs, V. R.: 1986, *The Health Economy*, Harvard University Press, Cambridge, Massachusetts, especially Chapters 15–17.
13. Hansen, A. S.: 1986, 'Cost-Containment', *Medical Benefits: The Medical Economic Digest* **3** (3) (February 15), 1–2.
14. Heilbroner, R. L.: 1986, *The Nature and Logic of Capitalism*, W. W. Norton & Co., New York.
15. Himmelstein, D. and Woolhandler, S.: 1986, 'Cost Without Benefit: Administrative Waste in U.S. Health Care', *The New England Journal of Medicine* **314** (7) (February 13), 441–445.
16. Horvath, B., Pecci, J., and Gay, W.: 1985, 'Fewer Tests May Cost More', *The New England Journal of Medicine* **312** (25) (June 20), 1645–1646.
17. Iglehart, J. K.: 1985, 'Medical Care of the Poor – A Growing Problem', *The New England Journal of Medicine* **313** (1) (July 4), 59–63.
18. Leaf, A.: 1984, 'The Doctor's Dilemma – and Society's Too', *The New England Journal of Medicine* **310** (11) (March 15), 718–721.
19. Levinsky, N. G.: 1984, 'The Doctor's Master', *The New England Journal of Medicine* **311** (24) (December 13), 1573–1575.
20. Levit, K. R. *et al.*: 1985, 'National Health Expenditures, 1984', *Health Care Financing Review* **7** (1), 1–35.
21. Light, D. W. and Widman, M.: 1985, 'Letter', *The New England Journal of Medicine* **313** (24) (December 12), 1548.
22. Maloney, J. V. and Reemtsma, K.: 1985, 'Cost Containment by Naval Armada', *The New England Journal of Medicine* **312** (26) (June 27), 1713–1714.
23. Marcus, S. A.: 1984, 'Trade Unionism for Doctors: An Idea Whose Time Has Come', *The New England Journal of Medicine* **311** (23) (December 6), 1508–1511.

24. Massey, R. U.: 1986, 'Reflections on Medicine: Right Thing, Wrong Reason', *Connecticut Medicine* **50** (2), 133.
25. Massey, R. U.: 1986, 'Reflections on Medicine: The Meanest Sayings', *Connecticut Medicine* **50** (3), 207.
26. Mayer, T. R. and Mayer, G. G.: 1985, 'HMOs: Origins and Development', *The New England Journal of Medicine* **312** (9) (February 28), 590–594.
27. Mechanic, D.: 1985, 'Public Perceptions of Medicine', *The New England Journal of Medicine* **312** (3) (January 17), 181–183.
28. Mechanic, D.: 1985, 'Physicians and Patients in Transition', *Hastings Center Report* **15** (6), 9–12.
29. Norstrand, I. F.: 1986, 'Takeover of the Medical Profession by Nonphysicians', *The New England Journal of Medicine* **314** (3) (February 6), 390.
30. Nutter, D. O.: 1984, 'Access to Care and Evolution of Corporate, For-Profit Medicine', *The New England Journal of Medicine* **311** (14) (October 4), 917–919.
31. Percival, T.: 1975, *Medical Ethics, or a Code of Institutes and Precepts, Adapted for the Professional Conduct of Physicians and Surgeons* (1803), Robert E. Krieger Publ. Co., Huntington, New York.
32. Pertschuk, M.: 1981, 'The FTC and Health Care: The Role of Competition', *Connecticut Medicine* **45** (4), 238–240.
33. Proxmire, S. W.: 1986, 'Developing an Effective Health Care Cost Containment Strategy for the Nineties', *Congressional Record—Senate* (March 10), S-2328 – S-2331.
34. Rabin, M. T.: 1983, 'Control of Health-Care Costs: Targeting and Coordinating the Economic Incentives', *The New England Journal of Medicine* **309** (16) (October 20), 982–984.
35. Reading, A.: 1985, 'Involvement of Proprietary Chains in Academic Health Centers', *The New England Journal of Medicine* **313** (3) (July 18), 194–197.
36. Relman, A. S.: 1984, 'The Rand Health Insurance Study: Is Cost Shaving Dangerous to Your Health?', *The New England Journal of Medicine* **310** (23) (June 7), 1453.
37. Relman, A. S.: 1980, 'The New Medical-Industrial Complex', *The New England Journal of Medicine* **303** (17) (October 23), 963–970.
38. Salmon, J. W.: 1985, 'Profit and Health Care: Trends in Corporatization and Proprietization', *International Journal of Health Services* **15** (3), 395–418.
39. Schwartz, W. B. and Aaron, H. J.: 1984, 'Rationing Hospital Care: Lessons From Britain', *The New England Journal of Medicine* **310** (1) (June 5), 52–56.
40. Smith, A.: 1965, *An Inquiry into the Nature and Causes of the Wealth of Nations* (1776), Modern Library, New York.
41. Spivey, B. E.: 1984, 'The Relation Between Hospital Management and Medical Staff Under a Prospective-Payment System', *The New England Journal of Medicine* **310** (15) (April 12), 984–986.
42. Starr, P.: 1982, *The Social Transformation of American Medicine*, Basic Books, Inc., New York, N.Y.
43. Thurow, L. C.: 1984, 'Learning to Say "No"', *The New England Journal of Medicine* **311** (24) (December 13), 1569–1572.
44. Wald, M. L.: 1986, 'Doctors Weigh Strike Over [Malpractice] Insurance', *The Times*, New York (March 9).
45. Whalan, W. A.: 1986, 'Rationing and Free Choice', *Connecticut Medicine* **50** (2), 131.

WENDY K. MARINER

SOCIAL GOALS AND DOCTORS' ROLES: COMMENTARY ON THE ESSAYS OF ROBERT M. COOK-DEEGAN AND STUART F. SPICKER

The metaphor of physician as captain of the ship seems to assume a single or unified function. Most of the papers in this volume present evidence to the contrary, setting forth a panoply of functions, many of them irreconcilable, the primacy of which depends on the perspective of the observer. Indeed, physicians may serve as scientific investigators, diagnosticians, patient advisors, skilled technicians, team supervisors, palliative caregivers, consultants to legislators and policy makers, rationing agents, administrators, and more. This is no surprise.

What does surprise this writer is the rarity of discourse relating these various functions to the structure of the health care system in which physicians are expected to perform. For the social role of physicians in general (and their particular roles as individuals) varies with the goals society sets for the health care system as a whole. I shall argue that much of the ambivalence or disagreement surrounding the physician's proper role in the health care system may be attributable to the multiplicity of available roles (more particularly, to the existence of two distinct categories of roles) and to the failure of society to integrate effectively the goals of the health care system itself.

GOALS OF THE HEALTH CARE SYSTEM: ZIGZAGGING BETWEEN STRUCTURE AND OUTCOME

The United States health care system has three broad goals: efficiency, in the sense of economic efficiency and administrative simplicity; effectiveness, with respect to the quality of care and improved health outcomes; and equity, in the sense of an equitable distribution of services, with the assumption that access to some level of basic care be ensured. I shall not bother to justify these goals here. Rather, I take them to be descriptive of the overall objectives accepted by virtually all members of society. Indeed, the problem I address has nothing to do with the definition of such goals (although their definition is not without controversy), but with the paradox that, having posited or at least accepted such goals, we seem unable to keep them in mind all at once.

Nancy M. P. King, Larry R. Churchill, and Alan W. Cross (eds.)
The Physician as Captain of the Ship: A Critical Reappraisal, 177–187.

We have tended to pursue these goals one at a time, directing all our energies toward one until we discover (or remember) the relevance or necessity of another. Then we concentrate wholeheartedly on a different goal, almost ignoring the other two, until the cycle is repeated. In the early half of this century, most attention to the health care system concentrated on the goal of effectiveness – improving the quality of care, fostering research, finding new ways to prevent or cure illness [30]. In the 1960s and 1970s, we "discovered" that the fruits of these endeavors were not reaching everyone; improved national health could not be achieved unless all those in need had access to the benefits of health care. So we shifted the focus to equity, and sought to enable larger numbers of people to participate [6, 7, 17, 30, 32]. Now efficiency is the predominant concern. Having recognized that it is expensive to maximize both quality and equity, and having lost some confidence in earlier definitions of quality, we have turned our attention to reducing or minimizing the costs of providing health care [9, 13, 28].

This kind of focus shifting reflects the tension between what we want the health care system to *accomplish* and how we want the system to be *structured*. On the one hand, there is a general consensus that all members of society should have access to quality health care on the basis of need [26]. On the other hand, there remains a strong preference for keeping the delivery of health care in the hands of private individuals and organizations who function with maximum freedom of choice [1, 20, 25, 33].

The goal of equity (and to a lesser degree, that of effectiveness or quality) grows out of a popular form of egalitarianism, based on the intuition that health care is of special importance – so important that it should be available to persons just because they need it to survive or to keep from deteriorating [2, 4, 22, 23, 24]. Under this thesis, access to basic care should not depend on one's ability to pay.

Yet, as Uwe Reinhardt has noted, we have tried to "extract an egalitarian distribution of health care from a delivery system still firmly grounded in libertarian principles" [27]. In an attempt to expand access to care, we superimposed Medicare and Medicaid upon an essentially private system of health care delivery. We still adhere to the individualistic, libertarian conception of personal services, which values individual liberty.

The current efforts at cost containment have a marked libertarian ring to them [18]. They rest on the assumption that free market competition can produce economic efficiency and curtail unnecessary expenditures for health care. Of course, no one is actually proposing to create a pure market economy for health care in the United States. Indeed, most advocates of increased competition admit that real competition will have to be constrained or supplemented in order to enable those who cannot buy their way in to have access to the system [11].

Recognition of the need to modify the market structure reflects persistent concern that the pursuit of economic efficiency will, if unrestrained, threaten progress toward equity and quality of care. Yet we have never really tried to align the structure of the system with all our goals. Rather, we have tinkered with structure and distributional goals independently, so that the structure becomes a drag on progress toward our goals. Surely this has helped keep us shifting from goal to goal without addressing their influence on one another. Is it possible to structure an efficient health care delivery system that maximizes freedom of choice and also achieves equity in the distribution of high quality care? Given their different orientations, as well as the importance of the structure of the health care delivery system, it is unlikely that one simple approach can serve all our goals at the same time. We will undoubtedly need a combinaton of approaches to attack the different goals.

DOCTORS' ROLES: ZIGZAGGING BETWEEN SCIENTIST AND SAMARITAN

A further complication of the zig-zag approach to health care goal setting shows up in the peculiarly dichotomous views of physicians' roles described by Dr. Cook-Deegan [3]. On the one hand, the physician has been seen, often with more nostalgia than insight, as the good samaritan, the beneficent caregiver. At the same time, the physician has become the technologist, the scientist with knowledge and skill sufficient to defeat illness. The point was originally Walsh McDermott's [21]; Cook-Deegan gives it an especially modern gloss.

If technology has enhanced medical capacity, it has also deprived medicine of a measure of its distinction. The moral authority of physicians as a profession entitled to lead the development and use of health care derived primarily, if not exclusively, from their role as samaritans. Jay Katz points out that medicine achieved respectability when it could claim a scientific basis for its knowledge, but consolidated its power as a profession by insisting that scientific knowledge could be put to use only by the exercise of clinical judgment, a matter of training and experience beyond the reach of uninitiated laypersons [16]. As the health care system diminishes this latter function, increasing its reliance on technology in the broad sense, the clinical role of the physician reduces to that of technician who has scant claim to any position of control over the distribution of resources.

The result is a dilution of the moral claim of physicians as a group to direct the shape of the health care system. This dilution of the moral authority of physicians in their social role is distinct from the division of power between individual physicians and individual patients. There the notion of moral authority has little to recommend it as against the autonomy of the patient [10, 16]. But the argument against physician dominance of decision making for individual patients, which can be based on patient autonomy alone, is very different from the argument against the moral authority of the physician to help control the development and distribution of medical technology and resources.

The loss of recognition of the samaritan function is important to the social role of physicians in the health care system for two reasons. First, it deprives physicians of an integral element of their professional definition in their clinical capacities, that of caregiver. One might argue that, empirically, physicians are caregivers and that clinicians, at least, perceive themselves as such, so that to deny or repress this aspect of their professional service is simply wrong as a matter of fact. More important, it deprives them of any claim to make decisions affecting health care policy. If they are mere technicians, without any special characteristics giving them an enlightened perspective on the health care system, they ought not to be made responsible for decisions affecting health care policy.

Insofar as the health care system experiences reforms and modifications as a result of focus shifting, whereby goals are pursued seriatim rather than simultaneously, the social role of physicians also moves from samaritan to technologist and back again. In the pursuit of quality, we expect physicians to provide the best for their patients, regardless of cost. In the pursuit of equity, we expected even more – that physicians provide it all to everyone. In the pursuit of efficiency, we seem to be saying that we cannot have it all, and that physicians will be expected to decide who gets what – sometimes as mere technicians applying rationing criteria according to scientifically determined need; sometimes as employees and bureaucrats applying health planners' rules; and sometimes as samaritans, given the full power – and onus – of fair and compassionate judgment.

It should not be surprising, then, that there is more confusion than clarity over what role physicians should play in the construction of the health care system. Some of the confusion appears to stem from our failure to recognize the utility and merit of having physicians in a variety of roles within a complex system that requires technicians and caregivers, as well as policymakers. Indeed, physicians may assume different roles at different levels of the system. Yet, in the absence of a clear concept of the structure and purposes of the system as a whole, it remains difficult to agree on the proper mix of roles for physicians and on whether the samaritan or the technician is better equipped to guide the system. It may only be possible to align physicians' roles if the overall goals of the health system are themselves aligned and integrated.

DEFINING ROLES WITHOUT DEFINING GOALS

While Robert Cook-Deegan emphasizes the variety of roles for physicians in a technological world, Stuart F. Spicker suggests a new role for physicians in a competitive health care management model [29]. Professor Spicker argues that physicians, managers, and perhaps patients should be able to structure private arrangements in a competitive environment that achieve all three social goals of equity, efficiency, and quality of care. He is right to challenge simplistic fears of introducing economic efficiency into health care delivery; still, he cannot rely on the good will of business to reconcile all three of our conflicting objectives. Fair competition has unfair results. Ensuring fair competition among competing providers will still leave a large proportion of the patient population out in the cold (although it might not hamper quality). Professor Spicker rightly points out that economic competition creates clear incentives to avoid providing care to the very sick and the poor, who are often the same. Some form of counterincentive designed to promote equity will be necessary to achieve any fair distribution of care in a competitively structured system. Competition can only be shaped to produce particular distributive results by the imposition of constraints on the free exercise of market choices; that is, by coercion. For example, health plans and other providers might be required to accept a certain proportion of their patients from among the poor and medically needy.

Professor Spicker is sensitive to the need to place constraints on the way the system is structured if we are to achieve our distributional goals. Yet he evidences some understandable ambivalence about what kinds of constraints are appropriate. He seems to assume that physicians and managers will act rightly, as they should. But why should we grant that? If it is important that people behave as they should, we must make sure that they do. The question is how? The economic incentives of our current essentially private health care system ignore or conflict with the goals of equity [19]. Ethical obligations of physicians are too personal to promote equity effectively across a national health care system. It seems unlikely that appropriate guidance can come from anything other than law, which is broad enough to cover the population affected and strong enough to create serious constraints of fairness on the delivery of care.

Professor Spicker too wonders whether the managers in a competitive system will create the kind of corporate culture that fosters the tripartite goals of quality, equity, and efficiency. He warns against too much reliance on altruistic conduct, suggesting instead that managers and physicians enter into contractual arrangements to keep managers from overstepping their mandate and competence and from intruding into the physician-patient relationship. But ultimately he concedes that other legal and regulatory mechanisms may be required to avoid possible failures resulting from excessive self-interest.

To develop such mechanisms, he distinguishes two "gatekeeping" roles for physicians. The first, in which the physician acts to ration resources for individual patients on economic grounds, he rejects as incompatible with the clinical responsibilities of physicians to individual patients. The second, however, makes the physician a necessary participant in *a priori* policy decisions respecting resource management. This I take to mean that physicians themselves are a valuable source of information in determining what kinds of services are worth providing in particular institutional settings and should be involved to enlighten choices that might otherwise be made on the basis of erroneous economic assumptions about optimal standards of care. In short, Spicker is here endorsing the physician's samaritan role at the policymaking level.

The result of these policy determinations, according to Spicker, should be to avoid the tragedy of the commons. He is right to urge that physicians distance themselves from economic gatekeeping for particular patients [5]. But it is the second gatekeeping function that determines what is available to patients. It is unclear whether private decisions made independently at the level of health care institutions can achieve the broader goals of the nation as a whole. If we add up the decisions of thousands of organizations, we may find that we do not like what they decide to make available, or that whatever is available is not available on fair terms.

That is to say that the structure of our health care system may not further what we want the system to accomplish. In theory, perhaps, a fair configuration could be developed by private agreements among physicians, managers, and patients. Spicker is right to say that we cannot yet claim that private decisionmaking will necessarily lead to bankruptcy, greed, or tragedy. But he underestimates the difficulty of achieving a proper balance of equity, effectiveness, and efficiency because he fails to address the relationship between goals and structure or that between systemic goals and physicians' roles. Most importantly, he fails to insist that goal definition must precede role definition and contractual specifications before any of these can be said to work. It seems unlikely that fairness-producing private contracts *will* be created by the free will of the parties in the absence of some fairly strong externally imposed constraints. By the same token, neither will we know whether decentralization and competition are successful structural principles until we know precisely what it is we wish to achieve through them and how those achievements relate to our three systemic goals.

FINANCIAL INCENTIVES WITHOUT GOAL DEFINITION

If we wish to restrict the amount of national health care expenditures in the service of efficiency, we must reduce the *size* of the commons (or what is included in the commons) or limit the total *use* of the commons. Current financing mechanisms based on prospective payment are directed toward limiting the use of the commons. Perhaps we assume that this will reduce its overall size; in reality, it is more likely to change the *way* it is used, and the people who use it – in short, to keep some people out of the commons entirely. We have not yet defined the size of the commons, what it must include, what it may include if resources permit, or who should be allowed in on what terms. We have not yet attempted, that is, to determine how this means of pursuing efficiency in care affects the pursuit of effectiveness and equity – or even to ask how the latter two should be pursued within the current structure of caregiving.

We are keeping the physician in the nominal role of gatekeeper, but we have changed the gatekeeper's real function. In the past the gatekeeper decided what patients would receive from among what was perceived as an unlimited commons. Now, we are asking the gatekeepers to decide what patients will receive from a limited commons [8, 15], but we won't tell them how the commons is limited.

Physicians are asked to make explicit policy choices because no such choices follow from the system's structure. Prospective payment systems, like Diagnosis Related Groups (DRGs), are supposed to provide the clues – the incentives – to prompt physicians to open the gates only when it serves efficiency. Within such payment systems, there are no real financial incentives fostering access to care on the basis of need alone and precious few fostering effectiveness or quality of care [31].

Part of the problem with using payment methods as incentives for determining the use of the medical commons is that they are based firmly in the fee-for-service structure. The fee may be lower now, but it is still a per unit payment. This structure offers few clues as to how to serve our goals – what the service *should* be or *who* it should be provided to. The economic incentives, rather, are to provide as many units of service as possible, and as little service per unit as possible. Thus, the commons may suffer as much or more use as before, and may serve more people, but it remains unclear whether the users (much less the non-users) are getting what they need.

We have not decided *what* services we as a society want to pay for. Instead, we are worrying about which service configuration to use as the basis for calculating payment [12, 14]; but payment calculations do not answer the question of what services are worth paying for. Deciding what services are worth paying for is part of defining what is to be included in the health care commons. Such decisions obviously depend on the purposes the commons is to serve.

This returns us to the need for making explicit the goals of the health care system or, more particularly, the balance among the three goals of quality, efficiency, and equity. The fact that we have failed to do so thus far attests to the complexity of the task. Indeed, it may be that we hoped to avoid facing the difficulties of specifying what is and is not necessary by shifting the decision making from the national level to the level of private institutions like hospitals and health maintenance organizations in the hope that their individual actions would somehow achieve the balance that we were not able to specify. Our preference for maintaining a privately organized structure for the delivery of care has kept us as a society from confronting the need for deciding what the structure should accomplish.

CONCLUSION

Confusion and frustration over the proper definition of the role of physicians in today's health care system may be symptomatic of our failure to integrate and balance the goals of the system as a whole. If efficiency were all that we sought, we would experience few qualms about physicians who act as economic rationing agents. By making explicit the need to make certain kinds of health care available to all persons and the limits of the financial resources to support such care, we can, by an iterative process, come to a definition of the kinds of care that can be provided in different contexts. This need not entail extensive national control over the actual provision of services. As long as the overall goals are specified, it should be possible to permit a mix of institutions and services to achieve them. More important, it will be possible to determine whether private decisions are furthering or hindering the success of the system. In particular, we can then assess the value of various technologies in relation to our goals, and decide either (1) what technologies should be permitted in the commons, or (2) what technologies we are willing to pay for (perhaps leaving others available on the basis of ability to pay). The first purpose involves the more direct command and control form of regulation, which prohibits in one way or another the use of technologies found to be ineffective or unnecessary. The second involves financial incentives to ensure that public funds are spent where they do the most good and to redirect private choices to effective services.

With the goals of the system made explicit, it will also be possible to define the role of physicians with greater clarity and less conflict. In fact, it is likely that there will be many roles for physicians. But the difference is that they should be publicly announced; that is, there should be no covert incorporation of economic roles for physicians who are making decisions about the services necessary for an individual patient. Rather, other physicians who occupy public positions in the government will contribute to decisions concerning the efficiency of certain kinds of services. There does not appear to be any need to force all physicians into one role or the other, be it samaritan or technician or economic gatekeeper. Nor do we need to expect physicians to leapfrog from one role to another and back again. Different physicians will perform different functions for specific purposes.

If the medical commons can be thought of as navigable waters, it will support a mixed fleet, with some physicians navigating supertankers, others on supply boats and tugs, and still others in the coast guard monitoring the traffic. There are different roles for different goals. But first, we must recognize what our goals are, and we must acknowledge that they are not all likely to be achieved at the same time without specific direction from government – direction in the form of legislation guaranteeing access and promoting the quality of care as well as limiting our health care expenditures.

Boston University
Boston, Massachusetts

BIBLIOGRAPHY

1. Blumstein, J. F. and Sloan, F. A.: 1981, 'Redefining the Government's Role in Health Care: Is a Dose of Competition What the Doctor Should Order?' *Vanderbilt Law Review* **34**, 849–926.
2. Buchanan, A.: 1984, 'The Right to a Decent Minimum of Health Care', *Philosophy & Public Affairs*, **13**, 55–78.
3. Cook-Deegan, R.: 1988, 'The Physician and Technological Change', in this volume, pp. 125–158.
4. Daniels, N.: 1985, *Just Health Care*, Cambridge University Press, Cambridge.
5. Daniels, N.: 1986, 'Why Saying No to Patients in the United States Is So Hard: Cost Containment, Justice, and Provider Autonomy', *The New England Journal of Medicine* **314**, 1380–1383.
6. Davis, Gold and Makuc: 1981, 'Access to Health Care for the Poor: Does the Gap Remain?' *Annual Review of Public Health* **2**, 159–182.
7. Davis, K. and Schoen, C.: 1978, *Health and.the War on Poverty*, The Brookings Institution, Washington, D.C.
8. Eisenberg, J.: 1985, 'The Internist as Gatekeeper: Preparing the General Internist for a New Role', *Annals of Internal Medicine* **102**, 537–543.
9. Enthoven, A. C.: 1980, *Health Plan: The Only Practical Solution to the Soaring Cost of Medical Care*, Addison-Wesley, Reading, MA.
10. Faden, R. R. and Beauchamp, T. L.: 1986, *A History and Theory of Informed Consent*, Oxford University Press, New York.
11. Fried, C.: 1976, 'Equality and Rights in Medical Care', *Hastings Center Report* **6**, 29–34.
12. Hadley, J.: 1984, 'How Should Medicare Pay Physicians?' *Milbank Memorial Fund Quarterly* **62**, 279–299.
13. Havighurst, C. C.: 1977, 'Controlling Health Care Costs: Strengthening the Private Sector's Hand', *Journal of Health Politics, Policy & Law* **1**, 471–498.

14. Hornbrook,: 1983, 'Allocative Medicine: Efficiency, Disease Severity, and the Payment Mechanism', *Annals of the American Academy of Political & Social Science* **468**, 12–38.
15. Kapp, M. B.: 1984, 'Legal and Ethical Implications of Health Care Reimbursement by Diagnosis Related Groups', *Law, Medicine & Health Care* **12**, 245–253.
16. Katz, J.: 1984, *The Silent World of Doctor and Patient*, Free Press, New York.
17. Lewis, C. E., Fein, R. and Mechanic, D.: 1976, *A Right to Health: The Problem of Access to Primary Medical Care*, Wiley, New York.
18. Mariner, W. K.: 1983, 'Market Theory and Moral Theory in Health Policy', *Theoretical Medicine* **4**, 143–153.
19. Mariner, W. K.: 1987, 'Prospective Payment for Hospital Services: Social Responsibility and the Limits of Legal Standards', *Cumberland Law Review* **17**(2), 379–415.
20. McClure, W.: 1982, 'Implementing A Competitive Medical Care System Through Public Policy', *Journal of Health Politics, Policy & Law* **7**, 2–44.
21. McDermott, W.: 'General Medical Care: Identification and Analysis of Alternative Approaches', *John Hopkins Medical Journal* **135**, 292–321.
22. Moskop, J.: 1983, 'Rawlsian Justice and Human Right to Health Care', *Philosophy & Public Affairs* **8**, 329–338.
23. Outka, G.: 1974, 'Social Justice and Equal Access to Health Care', *Journal of Religious Ethics* **2**, 11–32.
24. Ozar, D. T.: 1983, 'What Should Count As Basic Health Care?' *Theoretical Medicine* **4**, 129–141.
25. Pollard, M.: 1983, 'Competition or Regulation: A Critical Choice for Organized Medicine', *Journal of the American Medical Association* **249**, 1860–1863.
26. The President's Commission for the Study of Ethical Problems in Medicine and Biomedical and Behavioral Research: 1983, *Securing Access to Health Care*, U.S. Government Printing Office, Washington, D.C.
27. Reinhardt, U.: 1985, 'Uncompensated Care: Legal and Social Perspectives', *Hospital Law Newsletter* **2**, 1, 5.
28. Schweiker, R.: 1982, *Report to Congress: Hospital Perspectives Payment for Medicare*.
29. Spicker, S. F.: 1988, 'Marketing Health Care: Ethical Challenge to Physicians', in this volume, pp. 159–176.
30. Starr, P.: 1982, *The Social Transformation of American Medicine*, Basic Books, New York.
31. Stern, R. S. and Epstein A. M.: 1985, 'Institutional Responses to Prospective Payment Based on Diagnosis-Related Groups: Implications for Cost, Quality, and Access', *New England Journal of Medicine* **312**, 621–627.
32. Stevens, R. and Stevens, R.: 1974, *Welfare Medicine in America: A Case Study of Medicaid*, Free Press, New York.
33. Stockman, D.: 1981, 'Premises for a Medical Marketplace: A Neoconservative's Vision of How to Transform the Health System', *Health Affairs* **1**, 5–18.

SECTION IV

CAPTAINS, COMMITTEES, AND COMMUNITIES

RAYMOND S. DUFF

UNSHARED AND SHARED DECISION MAKING: REFLECTIONS ON HELPLESSNESS AND HEALING

In this essay, I will (a) describe some notions about caring and about the central position of what I call "moral community" in deciding and providing care of sick persons; (b) sketch some aspects of the evolution of the role of the modern physician as "captain of the ship" – the ship being hospitals, doctors' offices, homes, and the people who apply science, technology, and other means to care for the sick; and (c) comment on the role of hospital ethics committees.

In recent times, hospital ethics committees have arisen as part of the ship's bureaucracy, which has been expanding along with the rapid growth of technology and specialization in practice. The role of ethics committees is examined and discussed in the perspective of the best traditions of caring. I suggest that further major reforms of medical education and practice are needed. The Flexnerian reform did only half the job. The more difficult half is yet to come, and ethics committees are not up to this. In terms used in this volume, I assert that medicine needs a new synthesis; the captain and the public, a new orientation; and the ship, a management much more centered on patients and families.

CARING AND MORAL COMMUNITIES

The aims of caring are to help the sick get well, to reduce suffering, and to help well persons stay well. Caring, always complex and more or less unique for each person, involves a variety of technical, social, and personal means. Since true caring requires that the ends of care be shaped primarily by those cared for [43, 44], decisions of professionals must be shared with the sick and often their families. Without this guiding principle, caring necessarily will be blind. Unshared decision making in patient care fosters helplessness. Shared decision making fosters healing, a sense of wholeness much desired by most persons. These linked propositions underlie the whole argument I present here.

Throughout history, people sick or well have always felt the need for human love and caring in order to cope with threats which they know are all around them and, even worse, within them (because of disease

Nancy M. P. King, Larry R. Churchill, and Alan W. Cross (eds.)
The Physician as Captain of the Ship: A Critical Reappraisal, 191–221.

risks inherent in the human condition). In illness, feelings of need for care intensify because illness separates the sick from others and even themselves. Illness is a disintegrating experience. People feel torn apart [10, 59]. But suffering is eased, disability is less burdensome, and dying is more tolerable when the helping touch of compassionate hearts and hands is felt. Despite despair, a cherished sense of some wholeness can usually be protected. And wholeness or the feeling of wholeness is the prime aim of healing.

Most of us acknowledge that, because of social ties in living, caring for the sick is, within limits, also caring for the well who do the caring. Those who care feel good about themselves, for good reasons, most of us think. They are doing nice things for people. If they succeed dramatically, they achieve heroic status [54]. But there are risks in this. Caring may or may not help those who care to understand and cope with the basic existential facts of their own lives and the lives of others. If they have learned the lessons of their work well, they may live better than other people; if not, they may perform poorly in their personal and professional lives [28, 46].

Moral communities as I have seen them over the years consist of a patient, his or her intimate social others (usually family), and their advisors. There are as many moral communities in health institutions or ambulatory settings as there are patients seen in them. Functionally, moral communities represent decentralized decision making because choices are made by particular, unique persons and those close to them (family and professional advisors). Decisions will necessarily reflect what is most valued at the intimate level of living where, as Berlin noted, the most important aspects of life are created, celebrated, and transmitted ([4], p. 15). Of course, the advice of technical experts is important so that persons can make informed judgments about treatment recommendations. Illusions have to be avoided through an often slow process of truth telling. Magical beliefs can be replaced gradually, but perhaps only partially, by an understanding of reality. Painful though this may sometimes be, it is likely to be helpful and far less dangerous than illusion.

A high degree of freedom is implied in the function of moral communities, and surely there is some risk in that. Of course, one would hope that they would function responsibly, fairly, and wisely. Since none can be perfect, they will not always do that. But, within limits, they will have to be the final arbiters of these virtues. Clearly not everyone will agree with individual choices that are taken. By this model, they don't have to. It is important to note that the alternative to this strategy of decision making is most likely to be a set of rules that please a few and tryannize all others [48].

The usual order of decisional power regarding moral issues (that is, the *ends* of care) in moral communities is: patient, intimate other (usually family), and health advisors. Among health advisors, the physician may or may not be implicated in particular situations. Sometimes nurses, social workers, family members, or friends are more competent to decide particular questions such as use of sedatives, narcotics, or even resuscitation. I assume, of course, that whoever decides will have the advice and usually the consent of those who are relevant advisors.

The order of decisional power in carrying out decisions (which concerns the *means* of care) is often different. Here, technical experts must dominate some decisions. This situation is like building a highway through a city. Engineers can tell us how to do it best. But they cannot tell us whether building it is a good idea or, if so, where to put it.

In brief, moral communities give democracy a chance. They foster "horizontal" human relationships. Trust, education, and mutual respect are central values. *Physical*, *personal*, and *social* conditions are given thoughtful consideration *together* in deciding and providing care [43, 66]. Since decision making is decentralized, decisions have a good chance to be based on existing strengths of sick or troubled people – which is empowering and vital for coping with illness. This general approach to caring in moral communities is given compelling support by scientific literature on learned helplessness [62], which I will not review here except to say that a sense of control is necessary for the achievement of control and that a sense of no control usually ensures failure.

MORAL COMMUNITIES IN HISTORY

How have moral communities as I have described them fared in history? Not well. For more than 2000 years, medical practitioners have demonstrated remarkable paternalism and extensive authoritarianism that frustrate the functions of moral communities [11]. Of course, there have been notable exceptions to this. For example, Plato's physician for free men "treats their disease by going into things thoroughly from the beginning in a scientific way and takes the patient and his family into confidence. Thus he learns something from the sufferers, and at the same time instructs the invalid to the best of his powers. He does not give prescriptions until he has won the patient's support, and when he has done so, he steadily aims at producing complete restoration to health by persuading the sufferer into compliance" ([53], pp. 104–105).

This is a concise description of moral community and its proper function as I have described it. The parties are designated. Learning by all is emphasized. Decision making is shared. Feedback is constant. This scheme fits nicely with the strategy of decision making, evaluation, and "incremental" change proposed by Braybrooke and Lindblom for complex situations that among other things are characterized (like health care) by vast complexity and limitations of resources [7].

But Plato has always been counted among the world's arch elitists. He also described the role of physicians for slaves: "A physician who treats slaves never gives him any account of his complaints, nor asks him for any; he gives him some empiric injunction with an air of finished knowledge in the brusque fashion of a dictator, and then is off in hot haste to the next ailing slave . . ." ([53], p. 104).

In a lengthy review of physician relationships with clients, Chapman finds: "Over the past two millennia, the medical calling has approached the matter of its ethical *raison d'être* very timidly indeed" ([11], p. 145). Throughout this time, people of assorted healing beliefs tried to help the sick while making a living. They competed with one another. In general, client ignorance and passivity were advantageous for the healers. In the face of vast ignorance, there was little to discourage healers and clients alike from holding to "miracle, mystery, and authority," as the clergy has always been inclined to do. Ladd pointed out that healers and the clergy shared a common fondness for paternalism. In the twentieth century, however, physicians, increasingly scientific themselves, have become attracted to science and philosophy [34].

During the 50 years from 1870 to 1920, medicine in the United States was transformed from a powerless, disorganized, and diverse group into a scientific endeavor dominated by institutions in which biomedical research was "enthroned" and thus far remains so [37, 68]. Specialization in medical practice and fragmentation in health care became dominant. Generally, personal and social influences on disease and illness were placed at best in the background while a hierarchy of health professionals and institutions dominated the scenes of caring.

Mixed results followed from these changes. Morbidity and mortality rates declined, though more as a result of public health than as a result of treatment measures. People now live longer and better. It is unimaginable that people would want to return to the unmodified natural state of massive, uncontrolled suffering and high rates of premature mortality. But the means of this progress have impersonal and intrusive elements in them. These tend to force disintegration upon people. Suffering is commonly and knowingly induced by those who care – an irony surely not new in history but one that may be less visible and yet more pervasive than it was in the past. Healers (as always) act with benevolent intent but now they use impressive, controlling, and sometimes mysterious means (mysterious at least to the sick or their families), which not only bring great benefits to some but also may harm others while serving as buffers or barriers between people [56]. The mysteries of healing bring *attributed* but not necessarily *earned* power to the healer [54]. The hopes and illusions of the public and professions alike have always encouraged quackery [75]. Technology may serve to confuse, to keep people ignorant of one another's knowledge and deeper feelings. Then, where is our monitoring of our *real* power, our sense of caring, and the impact for better or worse of our attempts to heal? Lost, maybe. We will look more carefully at these issues.

Persons involved in the care of the sick have always included the patient, intimate others in the social environment, and health and other advisors. In the case of modern medicine, a vast technology has arisen. As a result, many more persons are involved in care. Their work and much of their relationships with patients have come under the control of the physician, the acknowledged superior expert about human bodies and about the use of technology to protect bodies. The physician has become known as "captain of the ship," in command of the people and the technology used to combat disease and death. The military metaphor has been and is used commonly. It draws attention to the basic existential problems and struggles of life: fear of suffering, loss, and death and battles against these. The physician regularly takes control of the body even though he may know very little about the person within that body or about intimate others in the person's social environment. (I will have more to say about the reasons for this presently.)

One of the consequences is an implied alliance between physicians, who focus chiefly on the body as a machine, and pro-life ideologists. The former tend to ignore personal and social issues in order to get on with treating ailing bodies. The latter in all doubtful situations (and doubt is nearly ubiquitous) find that personal and social issues are irrelevant in deciding the treatment of bodies when life is in danger. As a result, moral communities are threatened by physician-technologists and by ideologists who in effect may be co-conspirators against moral communities. Some physicians are paternalistic technologists-ideologists, an apparent example being the current Surgeon General [60].

In retrospect, it appears that the Flexnerian reform was only half a reform [22]. Flexner and others (see comments about Cabot and Codman below) knew about the need for more extensive reforms, but most of the central figures in shaping policies at that time opposed these. They already had their hands full, and they knew that practitioners and some professors resisted much of what they were already trying to do. Besides these considerations, there was in place a cherished, flourishing medical paternalism (see discussion of Osler's influence below), which was not likely to yield even to urgently needed reform.

But eventually, economic conditions increasingly prompted reform. In order for caring to be possible in the modern era, ample resources must be made available because the magnificent engines of science are expensive to build and operate. Failure to allocate substantial resources for care of the sick would constitute a tyranny for everyone. The tyranny of the well over the sick would be obvious. But even well persons would be tyrannized (rightly) by the guilt arising from their default, which surely could not escape them entirely. If people do not care, they cannot gain the personal satisfaction just mentioned; and they cannot expect care for themselves when they become ill or disabled. The need for resources to care for the sick is asserted again and again for understandable humanitarian reasons. It is asserted also because health professionals and institutions must pay their bills and make a living. In this situation, only a simpleton could overlook conflict of interest.

The public is usually willing, even eager sometimes, to pay generously because it desires to induce healers to prevent disease, suffering, and death, to make them well and keep them so. Often with an obvious sense of urgency, patients and families, in addition to desiring human caring, seem variously to want applications of science, magic, the science of magic, or the magic of science. (Someone with the skills of a Shakespeare is needed to give adequate voice to the variety and intensity of expressed needs or wants.) As a result, the cost in money and emotional drain is now so high that we must consider another part of reality. Namely, failure to recognize the limits of economic and human resources available to care for the sick constitutes a tyranny of the sick over the well *and* eventually, though subtly, a tyranny of the sick over everyone *including themselves*. This second tyranny usually is not stated explicitly because to do so seems, on its surface, chillingly noncaring. This recognition could be used, for purely selfish reasons, to withdraw care. But failure to recognize and deal with both sides of the resource allocation issue is really non-caring and eventually even absurd [69].

This issue is associated with another change. Because of delayed mortality in recent times, we now have reached a point of diminishing returns for added efforts [23]. This complicates decision making. In modern health care some choices bring great benefits and the costs are minimal. Such choices are easy to make. Other choices are more difficult because the costs are higher and the benefits are in doubt. In addition, doubt cannot be eliminated before a costly commitment must be made. Finally, there are truly tragic choices where someone must lose, perhaps heavily, no matter what choice is taken [9].

MORAL COMMUNITIES AND MODERN MEDICINE

Before physicians had the ability to treat diseases with some measure of success, paternalism may have had a significant benefit; and perhaps it brings some benefits today. If a patient feels cared for, that is an important component of healing, and there is always the placebo effect which may operate in addition to this. So I suggest we should be cautious in judging paternalism. But I do assert that the paternalistic tradition joined with the role of physician as technical-biological expert eased the way for the modern physician to become "captain of the ship," with more power over his clients than he ever had in history. Hence, it seems reasonable to review in greater detail the record of the profession in supporting or neglecting moral communities since the dawn of modern medicine. I will show that hierarchical relationships, generally detrimental to the functions of moral communities, tend to prevail over others.

Sensitive practitioners noticed early in the development of modern medicine that medical technology was intrusive and sometimes abusive. Moreover, it sometimes caused harm. Taking note of this, trustees and administrators of hospitals well into the twentieth century often were reluctant to allow medical students and researchers access to the sick ([37], pp. 152–155). In addition, relying on biomedical knowledge alone to treat bodies was rarely sufficient in the understanding of causation, pathogenesis, and prognosis and treatment of illness.

One of the physicians most sensitive about these issues in the early twentieth century was Richard Cabot, who believed that patients must be recognized as persons in the provision of all care [8]. He noted, for example, that lying and evasions were common in the relationships between practitioners and their clients and found that this practice could not be supported by the results he observed. Deceit, though with benevolent intent (typical of paternalists), had serious, detrimental short-term and long-term consequences. The difficulty here has been known since Plato invented the "noble lie" as a device to get underlings to do what was good for them – "good" being determined by higher authorities, the philosopher-kings of Plato's Republic. Trust is destroyed when the lie is discovered (and it cannot be hidden for long because even underlings have brains). Feelings of security are necessarily undermined, cooperation is compromised, and reliable feedback and education are reduced to nil. In brief, moral communities as previously discussed fall apart. Cabot's remedy was *team-work* of doctor and social worker and of doctor and patient. However, Cabot did not recommend explicitly that decision making be *shared* with the patient.

Cabot's proposal was almost entirely ignored except that his innovations with Ida Cannon helped to introduce social service into hospitals. But, even today, social workers and nurses in comparison to doctors are weak influences in the modern hospital [30].

Codman, a contemporary of Cabot, made heroic efforts to persuade hospitals to assess the end results of surgery in terms of the patient's function [13, 14]. He provided examples of this method from his own practice, and he tried to convince the trustees and the medical staff at Massachusetts General Hospital to adopt this method of assessment. He challenged leaders (physicians, administrators, and trustees) to evaluate physician performances in terms of "end results." Leaders among physicians would be appointed, based on these, and their appointments would continue only if their performances remained high.

About deficiencies in hospital care, Codman wrote, "At our charitable hospitals there is no one who dares make [major] criticisms at all. It is the duty of no one and it is in the interest of no one – except for the patients and the community" ([14], p. 63). In the front of his book, Codman wrote, "This volume is dedicated to Richard C. Cabot because I respect his motives, admire his courage and energy, but heartily disapprove of some of his opinions and methods, for he seems to want to reform the bottom of the profession, while I think the blame belongs at the top."

What Codman proposed might threaten physician-scientists because at that time the effectiveness of their approach to disease was more a matter of faith (part of the cultural belief system in the progressive era) than of established fact. The careers of physicians could be ruined if their performances were found wanting, as physicians knew was often the case (if they were honest about it). While Cabot challenged patients, physicians, and social workers individually, Codman challenged and threatened whole institutions and people in the high places. Moreover, he was a threat to the paternalistic authority of the profession as a whole. Thus, it is understandable that greater pariahdom fell on Codman. The medical staff rejected his proposal. The administration seemed bewildered by it. The trustees took no action.

Yet Codman persisted to the end of his life. His final act was to make preparations for the publication of a book setting forth his ideas [14]. After he died and by prior arrangement, this was printed and circulated, at the expense of his estate, to the medical staff and trustees of the Massachusetts General Hospital. But as in life, so in death his work was largely ignored.

Cabot and Codman noted the vast reforms of the Flexner era in regard to biomedical research. They approved of these reforms, but they felt strongly that research, teaching, and patient care based on the body as a machine were not sufficient for the care of the sick because people were far more than bodies. They shared the visions of the progressive era, but their vision was more comprehensive than that of Flexner. It constituted a challenge to paternalism and to physician dominance of patient care. Flexner and most professors in medical schools believed that the social and behavioral sciences were soft, not very important compared to biomedical science, which was their chief interest [73] and the source of their growing power.

In 1927, Peabody, taking note of these problems, published a lecture that ended dramatically in the claim: "[T]he secret of the care of the patient is in caring for the patient" [52]. Presumably, acting on this nugget of wisdom would provide some balance for technology, fragmentation, and specialization. Peabody's lecture has been quoted extensively, but there is no reason to believe that any systematic changes resulted from his influence. Although, as Shorter recently has shown, some physicians have always listened to troubled patients and their families and thus encouraged healing in the context of moral communities [65], such listening has not been a central part of medical education.

Osler, probably the most revered clinician of the twentieth century, more than he ever realized may have discouraged listening to patients and families [16]. He wrote mostly about diseases, in the French and English tradition of putting clinical observations together with laboratory and autopsy results. He wrote little about the role of patients, but others give us a picture of his attitudes and behavior toward them. One of his students wrote, "With the patients, who instinctively recognize him as a person of another order, he was sympathetic, cheery and jocose, usually, however, at our expense, or at the expense of human frailty at large. Then he often spoke in epigrams" [39]. Another wrote, "I remember that once, when Sir William could not get anything like a satisfactory history of a case, he turned away from the patient 'John' and said, 'you know it is sometimes a great advantage to have been a Vet. for you cannot be led astray by the history'" [42]. In a recent survey of patient and physician attitudes, Linn [36] found that "reassuring responses [of doctors] were the most preferred, followed by neutral, humorous, and hostile responses." In other words, humor as a primary professional response in physician-patient encounters is likely to backfire since it tends to trivialize illness and to demonstrate disrespect for the sick person.

Osler seems to have held a deep mistrust of feelings, particularly those of sick people. In his view, sick people were by nature extremely dependent and unreliable. Note the following remarks by Osler given to nurses: "Except in the warped judgment of the sick man, for which I have the warmest sympathy, but no respect . . ." and, referring to feelings and dependency, "that deep mysterious undercurrent of the emotions, which flows along silently in each of us, is apt to break out in the rapids, eddies and whirls of hysteria or neurasthenia. By a finely measured sympathy and a wise combination of affection with firmness, you gain the full confidence of one of these unfortunates, and become to her a rock of defense, to which she clings, and without which she feels adrift again" [50]. Osler believed that physicians should be "imperturbable," a virtue that could be best achieved by attaining excellence in diagnosing and treating diseases. In light of research into learned helplessness, I have to conclude that Osler (in addition to developing a much-needed, artful approach to disease) unknowingly gave us a prescription for dependency and depression of patients and families and for paternalism and "burn-out" in the medical profession.

Osler himself was tired, perhaps burned out, after about fifteen years at Johns Hopkins. He went to Oxford, where life was better for him. At the time of his departure from Hopkins, he explained his two primary ambitions to his colleagues: "to make myself a good clinician," and "to build up a great clinic on Teutonic lines, not on those previously followed here and in England, but on lines which have placed the scientific medicine of Germany in the forefront of the world" [74].

Ample support of what I take to be one of Osler's influences was given recently by Beeson in describing his own training 50 years ago. He candidly expressed embarrassment about the impersonal relationships between patients and professors and doctors in training. In the case of poor prognosis, physicians in training and professors alike commonly lied to patients and were rather smug about their skills in doing so [57]. In addition to this, Mishler [47] in some new research in linguistics found that in doctor-patient conversations the "voice of medicine" is supreme over the "voice of the lifeworld" of patients and families. To the degree of that supremacy, Mishler claims, medicine is "inhumane." (This is not to suggest that the "voice of medicine" is unimportant. It is only to assert that in the final analysis the "voice of the lifeworld" must determine the proper ends of the "voice of medicine.")

Let us pick up some more threads in this story.

In the 1930s, L. J. Henderson (a well-known chemist since the turn of the century) attempted to interest the faculty at Harvard Medical School in a scientific, systematic study of the patient and doctor as a social system [6, 29]. At that time, he was aging; he died in 1942. The medical school faculty was not interested in his ideas. When he became blunt about his perceptions of the ironies of modern medicine (specialization, fragmentation, and technological intrusions often carried out without regard for the patient's feelings), he, like Cabot and Codman before him, became a pariah among his peers at the medical school. However, there were some emerging behavioral scientists (Talcott Parsons among them) who took up Henderson's ideas [51].

In the 1950s, some innovations (for example, at Cornell, Western Reserve, and Colorado) in medical education were made in an attempt to teach medical students about personal and social influences on illness [35]. (I had the privilege of visiting all of these institutions while these programs were in progress.) As in the earlier Hopkins-Flexner reforms, the method of instruction involved direct, "hands-on" experiences of students. The idea was that intimate contacts of students in relation to patients and families might be as instructive about personal and social realities as had been a similar experience (begun in the Flexner era) in relation to the body as machine. However, as Morris has shown convincingly [49], approaching people as persons is vastly different from approaching physical bodies of persons. Applying Morris's notions to medicine, one can see clearly that what he calls "rituals" and "routines" in human relationships are most supportive of the role of doctors as dominant paternalistic technocrats.

Morris found that people as persons and the professionals who see them can really be understood in some situations (like crisis) *only* in "dramatic" encounters. Dramas have several features. There is *novelty*, at least for some of the actors; the activity or the outcome is *important*; the outcome is in *doubt*; and the actors feel they have some *control* over the outcome. (Note, as mentioned earlier, that control or at least the sense of control is *the* central issue in all of the research and clinical applications about helplessness.) But the innovations I have mentioned in medical education reflected little sense of drama and hence little concern for client control. In retrospect, it seems clear that the innovators, apparently loyal to the Osler tradition, were quite naive about these complexities. Moreover, they underestimated the pervasive influence of the well-established and by then well-funded biomedical researchers on medical school faculties. (I have already commented on the values of faculty.) These major institutional innovations suffered a fate much like that of Cabot and Codman.

Until the mid 1960s, the American Medical Association was successful for the most part in keeping government out of financing care of the sick [70]. But government leaders realized that support of endeavors related to health was popular. So they voted more and more funds for research, and of course the growing biomedical establishment gladly used the funds. Medical technology advanced; but more and more of it was "half-way technology" which was expensive and often not very helpful [72]. Control by medical technologists as captains on the ship was pervasive.

Several consequences followed from all this. Patients expected more and more, even unachievable perfection; and if they did not get it, they often sued in court [17]. Costs went higher and higher. Then, government entered the picture of paying the costs for a great deal of care

(Medicare and Medicaid). But, by this time, hospitals and the profession of medicine were increasingly troubled not only by issues raised and set aside in the Flexner era but by severe economic problems and by concerns about the fundamental purposes of medicine and how much should be spent to support health care or even life itself.

Since neither hospitals nor doctors had a tradition of consistently turning to their clients for help in resolving vexing questions such as I have just presented, usually they did what bureaucracies do best: continue operations as usual, avoid facing the problems as long as possible, and finally, if nothing else works, appoint a committee to study the problems. This approach has distinct advantages: useful, established functions can be continued, and evolutionary reforms are possible.

ETHICS COMMITTEES

In the first half of our current century, hospital committees of many kinds were created: tissue, pharmacy, laboratory, medical records, and others. In addition, committees were formed to deal with questions concerning sterilization and abortion. In the past 20 years, committees have been created to deal with problems of research concerning human subjects. Despite considerable fumbling in the operation of such committees, they have helped to clarify issues because they have been successful in developing criteria for making recommendations concerning some standards for patient care, reproductive choices, research procedures, and informed consent for participation in research. I will not deal further with these developments here.

Concerning tough questions like giving or not giving life-sustaining treatment and resource allocation, use of committees has been much more problematic. In 1975, Karen Teel, a pediatrician, suggested that ethics committees be used to deal with some vexing problems [71], and the Supreme Court of New Jersey in 1976 (referring to Teel's article) suggested in the Quinlan case that families and physicians facing very difficult choices should consult an ethics committee regarding prognosis. This ruling was odd because prognosis necessarily is based on medical, not ethical, analysis and because ethics committees did not exist in most hospitals. Moreover, even where such committees did exist, they were untested for the functions prescribed by the court. The New Jersey court seemed confused. It failed to note that well-grounded medical criteria for making decisions regarding use of life-sustaining treatment probably cannot be developed in a general sense (which is what law usually deals with), though diverse and unique criteria may be created instance by instance from the personal, philosophical, or religious views found in particular moral communities. The court seemed to recognize a problem here when it said that ethics committees should confine their function to establishing prognosis, but it failed to make clear why, then, an ethics rather than a prognosis committee should be used. I suggest the most likely explanation for all this is that the legal profession and the courts mistrust the public even more than does the medical profession. Katz helps us understand this issue, and he leaves us with little optimism [31].

Since the Quinlan decision has been referred to frequently and since several more or less similar cases subsequently have appeared in the courts and probably have had a pervasive influence on behavior, I will examine Dr. Teel's brief (four short pages) article in some detail. Her opinion is shared widely among doctors, and it was evidently given a great deal of weight by the Quinlan court.

Teel's paper is one of several published in the Winter, 1975 issue of the *Baylor Law Review*. That issue contains the papers representing diverse opinions that were presented in a forum concerning social and legal questions related to passive euthanasia sponsored by the State Bar of Texas and the *Baylor Law Review*.

At the beginning of her paper, Teel wrote:

> I must confess to feeling somewhat presumptuous in speaking under the title of "What the Law Should Be." This is anything but an easy question; if there is an altogether satisfactory answer, it would probably have been fairly obvious long ago and there would be little to consider at this time. I cannot really present to you any personal credentials in the field of medical ethics, forensic medicine, theology or in any of the other areas that might deal on a daily basis with problems to which we address ourselves today. I can, however, bring to you the perspective of a single practicing pediatrician who cares for infants and children and their families, and sees, with some regularity, the joys and tragedies that are part of caring for children, and who is all too familiar with the inadequate, unjust, and inhumane ways we often find ourselves dealing with the tragedies [referring here to severely handicapped children].

In this statement, Dr. Teel implicitly discounts the credentials of her profession, which from intimate acquaintance with numerous persons facing tragedy might have learned better than any other group in society diverse ways by which people may best deal with misfortune. For solutions she looks to others, experts presumably – not to the humanity of her clients, her profession, or herself, where it is most likely that the only specific, workable answers, in the final analysis, can be found.

Dr. Teel went on to deplore not operating on children with intestinal atresia when they had mongolism (note that this is strictly a biological criterion), but she made no reference to medical or lay opinion that this choice, though it always has a bad appearance, could be sometimes acceptable morally and medically by apparently thoughtful, careful people, both lay and professional [63]. She also objected strongly to the suffering associated with prolonged dying of children whose prognosis was hopeless (again, a strictly biological criterion). She asked whether we should "accelerate" dying. But she did not mention the strategies commonly used in the hospice movement and elsewhere. These movements, often with input by religious leaders, have found more or less acceptable ways of helping people to live well at the end of their lives – which also means to die well. Some people believe that these ways avoid the hazards so dreaded in any policy of active euthanasia and, more important, that they lead to the discovery and enhancement of sacred values in intimate social relationships. These values may guide and sustain throughout life. It may be that we can have the best grasp of living by learning from the dying and those close to them what it means to die well, which is the same as living well to the very end of life. Teel bypasses all this.

Teel believes that the rights of many children are not protected. But she does not question whether the meaning in the tragedies she discusses can be addressed in the language of rights. From thirty-five years of work as clinician and researcher into patient care issues, I am convinced it cannot be. But I know that millions do not agree with that. They simply have not thought it through.

Finally, in pressing for ethics committees, Teel implicitly turns away from a long tradition in medicine which, though paternalistic and elitist in many respects, has defended the place of the family and its health, religious, and other advisors in deciding the care of incompetent patients. Although Teel shows some ambivalence about the use of ethics committees when she says, "These are personal and private problems [and] we must preserve the latitude for individual and caring human judgement," she still presses for ethics committees as a way to solve the problems. This is contradictory.

The Quinlan court and others as well ignored the implicit and explicit limits and even the contradictions in Teel's misleading opinion. In my opinion, courts too often have hung a substantial part of their opinions on flimsy medical or social analyses. The result of this has been to set us all on a legalistic, more or less authoritarian campaign in which the focus of attention is more and more on *rights* to the best possible biological care, less and less on *responsibility* for dealing with tragedy in the context of moral communities. As Fuller [26] and Schoeman [61] point out, this fosters divisive individualism and forces cherished values concerned with "shared commitments" in moral communities to "sink out of sight."

Creating ethics committees became fashionable in hospitals. Such efforts were enhanced by the first nationwide conference on institutional ethics comittees sponsored by the American Society of Law and Medicine and held in Washington, D.C., in April of 1983. Most of those who presented papers at that conference showed practically no awareness of the history of medicine or of ethics committees I have given, and they displayed remarkable agreement about the value of ethics committees. The editors of the volume coming from this conference wrote in the introduction: "We believe that ethics committees can serve as a reasonable and valid institutional endeavor to increase understanding among all concerned – health care providers, families, patients, and society – as well as to resolve many of the ethical, legal, and medical dilemmas facing those who care for critically and terminally ill patients. That is the premise of this chapter, and the overwhelming conclusion of the chapters that follow" ([15], p. 6). This sounds more like an assembly of the converted and the curious than a conference intent on serious examination of problems deeply rooted in history, culture, and our souls. Examination of these problems will necessarily take a long time, and protracted controversy will be inevitable. There is no easy solution.

In the case of small communities and even some major medical centers, as one conference participant noted, "You often find good, practical wisdom, but you might instead find coercive pettiness." Apparently knowingly, he asked, "If you were a physician in a small community hospital, what would you do if the chairman of the hospital's ethics commitee told you: 'In the name of God, I forbid you to do so and so?' Or, 'I command you to do this or that?'" What person or whose God does the physician serve? There is ample historical documentation that the potential for conflict, confusion, and tyranny is enormous in this situation because, among other reasons, a quasi-legal authority like the hospital's ethics committee can easily meddle with deeply felt religious and spiritual values of the sick, their families, and those who try to help them.

TWO STORIES ABOUT COMMITTEES FACING CONTROVERSY

To illustrate some of the realistic constraints that limit the work of hospital committees facing controversy, I will relate two stories. The first concerns an attempt to study quality and cost of care. The second concerns an ethics committee.

With biomedical research paying off more and more, or at least so it seemed, demands of the public for the benefits of this reseach increased. I have already referred to escalating costs of care since the beginning of the scientific era. But what happened prior to 1950 was minor compared to what happened later. Since 1950 the percentage of gross national product devoted to health services has increased from around 4% to over 10%. These dramatic cost increases initially brought about demands for funding through insurance mechanisms. Private insurance carried part of the load until the mid-1960s when Medicare and Medicaid legislation was passed. This resulted in a vast increase in expenditures by the federal government for personal health care. One of the conditions for being reimbursed for services provided under Medicare and Medicaid was that hospitals were required to have a committee that examined utilization in an effort to keep costs within reasonable limits. The story of the utilization review committee is as follows.

In the mid-1960s, I had been doing some research coerning patient care which subsequently was reported in a book, *Sickness and Society* [18]. The chief of our medical staff was aware of the general nature of the research that I was doing with sociologist August Hollingshead. He felt that I would be a good choice to chair a utilization review committee that was to be responsible to the medical board of the hospital. As in most medical centers, the medical board consisted primarily of department chairpersons.

Taking the rhetoric of the law at face value, the chief explained his hope of putting in place a plan for the study of quality of patient care along with utilization and hence cost. He talked about attempts by the American College of Surgeons in the 1920s to evaluate patient care in American hospitals. He told me that much of the report that had been written was suppressed because of fear that releasing it to the public would result in closing some hospitals (like all inferior and many mediocre medical schools had been closed after release of the Flexner report in 1910). This would drive patients into worse circumstances: care by quacks. Athough he expressed disappointment with the decision of the ACS, he also said that in light of the long history of conflict with quacks, the choice may have been wise. (Note that the chief shared a low regard for the public's ability to judge matters for itself – a sentiment Beeson believed characterized the feelings of many physicians trained in these earlier times [57].)

I told the chief that I was interested in accepting this assignment provided that he and the chairmen of the clinical departments understood and supported the idea that quality of care would be studied along with costs. This condition was accepted, and a committee was formed with representation from each of the clinical departments and from administration.

In committee deliberations, it soon became clear that data from doctors, patients, families, nurses, and social workers would have to be acquired systematically if we were to understand the quality of patient care. This was so because our collective vision (at least in rhetoric) of illness and of patient care included consideration of physical, personal, and social data. The study, then, would probably include examination of performances both at the "bottom" (where Cabot thought the problems were) and the "top" (where Codman thought they were).

Resistance to this effort emerged immediately when the doctors discovered that they, their patients, and their relationships with one another would be scrutinized. Over a period of several months, attempts were made to negotiate differences. But these failed. (It has always mystified me how doctors could claim that the quality of patient care depended, among other things, on the doctor-patient relationship, and at the same time claim that a study of that relationship was not necessary in a study of the quality of patient care.) My position was that the committee had faithfully reviewed its charge and offered a proposal that would fulfill its mission. We would study the quality of care as well as issues of cost. The position of the medical board was that it had not intended to study quality in the sense that we had thought of it. They were interested in studying "patterns" of care as reflected from examination of the medical records. But I pointed out that the committee believed the medical record was a hopelessly deficient source of information for the intended purposes.

At a crucial time during the negotiations, an administrator told me: "Good and urgently needed as it is, the committee's proposal won't fly. Although the proposal was created by subordinates of respective department chairmen, it is clear that most of these department chairmen don't want this study done. Their appointees to the committee have no power to make it fly. And no one else has the power either." (That should dispel most doubt about the hierarchical nature of the hospital.)

I then told the chief that I felt the medical board should either alter its charge or offer an alternative which might accomplish his mission. After careful review of the situation, the chief told me, "The only thing wrong with your proposal is that it is at least twenty years ahead of its time, and I have to keep this ship on course." Accordingly, he suggested that I make compromises or resign. I resigned.

The outcome was that another professor who was interested in patterns of patient care was appointed chairman of the committee. He was pleased to do this work. But, to the best of my knowledge, most members of the medical staff felt that while the publications arising from this work confirmed some utilization patterns which had been expected, they cast minimal light on quality of care. However, these publications did help to set the stage for a subsequent innovation: the use of "diagnosis related groups" (DRGs) – a biologic criterion – as a guide for reimbursement in a cost control effort. Thus, from the standpoint of policy, money flow was now guided by images of requirements of the care of bodies as machines. What patients, families, and health professionals find is needed by other criteria is mostly irrelevant, less important than ever. I know that this was not the course the chief had in mind for his ship. But that is what we all got from the *outside* because *internally* we had rejected an obviously necessary, responsible alternative.

In brief, when the doctors defended their claim to control of their work on the basis of biological criteria, those who held the purse strings used the same criteria in setting reimbursement policies. But since biological criteria are insufficient for understanding patient, family, and professional needs and problems, all persons in these central roles have lost heavily: lost their autonomy and much of their capacity to be creative where creativity is vital.

As I tell this story now, I find my perception of it is different from what it was in those years. It is now obvious to me that the chief and William Osler (mentioned above) had much in common. The chief admired Osler. Both men were called "the chief" and were excellent clinicians who displayed deep and enduring loyalties to the profession of medicine. They were popular among practitioners because they were good teachers about disease and because they were reluctant to fault colleagues. They encouraged the growth of biomedical science in medicine, and that made them popular among professors; but neither made major contributions to science. In relationships with patients and families, they practiced, taught, and projected an image of "imperturbability" (Osler's term). From my point of view, these men had the opportunity to serve as a bridge between scientists and the public and between practitioners and patients. But, since they placed little trust in patients or the public, they served primarily as bridges between scientists and practitioners. In subtle ways, they were in harmony with the earlier code of ethics of the American Medical Association. They were paternalistic at its best (". . . unite tenderness with firmness . . .") and its worst (". . . and condescension with authority . . .") [12]. Most of all, however, they were politicians, corporate captains who had a primary responsibility for running the ship. This involved, among other things, accommodating powerful groups of doctors having diverse interests but agreeing on the central role of the doctor in deciding the care of patients.

I should point out here that in light of the historical and social settings of the roles of the chief and the department chairmen (particularly the ways they were taught robust confidence in themselves and in science – as they defined it), it is easy to understand the reasons why they behaved as they did. As I have pointed out above, they were not alone in their attitudes. Indeed, their attitudes were general. These conditions brought about a system of "hierarchical regionalism," a term used by Fox [24] to indicate typical beliefs and practices in the United States and Britain since 1910. According to Fox, it has been widely believed that health problems can be solved best by creating knowledge in laboratories of major medical centers and then applying that knowledge widely by encouraging increasing access to the newest drugs, surgeries, and so on. Information flow is downward from the laboratories to public – thus the term "hierarchical." Little or no attention is given to information flow in the opposite direction. In this belief system, both health professionals and their clients hold that client beliefs need not be taken seriously.

I should also point out that the committee and I were not alone in our views. Nurses, social workers, clergy, and hospital administration supported us. For example, the executive director of the hospital (also at that time President of the American Hospital Association) was most sensitive to the complexities of patient care, family concerns, costs, and outcome. He often discussed his thoughts about a photograph he had prominently displayed in his office. It showed a thin, dejected, stooped man walking alone away from a large modern hospital in the background. A caption read, "DISCHARGED, CURED." The foresight of this administrator about economics was shown as he asked again and again, "Isn't anyone worried about this progressive, relentless increase in cost of medical and hospital care?"

I have told this story to give you some hint of the nature of professional and institutional power, the way bureaucracies work, the "usefulness" of one committee, and what happens when principle is in conflict with power. You can find some of the same trend in a report by Bloom [5] and in the next story which concerns an ethics committee.

THE SECOND STORY

In 1968, *Sickness and Society* was published [18]. It created some furor for several reasons. The methods were less than perfect, and some of the findings may have been impressionistic or even simply reflections of the prejudices of the researchers. Some felt that the whole study was a misguided piece of muckracking [3]. But there were some findings in that study which no one disputed. In dealing with poor prognosis, relationships between health professionals on one hand and patients and families on the other were characterized by evasions, half truths, and lies. Nor was there any doubt that some of the unfortunate consequences in patient care resulted directly from these distortions or failures in communications. One of the results was a drive in Connecticut to establish hospice in this country. Another consequence was that I undertook a study with Dr. Campbell of the decisions we and others made in the care of severely defective newborn infants. In our clinical roles, we took seriously the idea of moral community noted above. When that study was published in 1973, substantial conflict arose and has continued to the present day [19, 21].

Out of these experiences, my department chairman felt that I would be an appropriate person to chair a committee having to do with ethics for the department of pediatrics. (The New Jersey court in the Quinlan tragedy – see above – had recently suggested that hospitals create ethics committees to help resolve some issues in patient care.) His charge was to draw up some general guidelines about deciding the care of infants and children who had a poor prognosis. We agreed on who should serve on the committee: doctors, nurses, clergymen, social workers, administrators, and an attorney or so. The committee met regularly for several months and found that agreement on some issues was impossible. But it was able to agree on procedural guidelines for deciding care. We acknowledged that by prognosis there was already reasonable agreement about most treatment decisions. Most children had a reasonably good future and were treated with full vigor. A few were dying and were made comfortable. A few others had a poor outlook for life or quality of life; and for these children we acknowledged that reasonable people reached different conclusions about treatment or survival. The committee proposed that authority for deciding care in doubtful situations be vested primarily in carefully and fully informed family members and children (when able). We felt that this was in keeping with the best of medical traditions from Plato's physicians for "free men" to the notion of moral community already presented. If there was conflict, we proposed procedures for resolving it. Any involved party was to be free to seek consultation with anyone as long as accepted customs of confidentiality were observed. In brief, these guidelines proposed that people simply do the work that they were presumed to be doing in the first place. Neither the committee nor anyone else concerned with it found good reasons for establishing an ongoing ethics committee.

From committee deliberations, we learned that one of the chief causes of confusion in a rapidly changing situation was that the identity of the child's physician at times was not clear. This might result from the involvement of multiple physicians, sudden and dramatic changes in the child's condition, and so on. Again, we recommended procedures, this time for ensuring that from the time of admission to discharge or death the child's physician would be clearly designated.

These guidelines were accepted by the chairman and were referred to from time to time as necessary. They were subsequently published [20] and used by many hospitals besides our own.

But that was not the end of the matter. These guidelines were not accepted by the hospital as a whole, and, as I'm sure you all know, the "pro-life" movement gained momentum over these years. Persons concerned with that movement locally and nationally were not satisfied to let moral communities as I have defined them deal with their own tragedies in more or less diverse ways, within apparently reasonable limits. As a result, our hospital like all others throughout the land came under increasing pressure from the federal government and other parties to comply with rigid rules for deciding the care of defective newborns. When our own institution was investigated (as it was, for we had published the results of our decisions), numerous people from within our institution and from elsewhere (a notable example being the Catholic Archbishop of Hartford) expressed dissatisfaction with the inevitable ambiguities contained within our guidelines. (You may recall that when moral communities address issues, some degree of freedom must be exercised by them.) Various persons representing religious, political, moral or medical persuasions were intolerant of ambiguity, and they believed that freedom to decide would surely encourage great evil. They demostrated again and again an intense drive to "escape from freedom," a widespread phenomenon described by Fromm [25].

I have reasons to believe that (as happened in our own hospital) more rigid guidelines that would give hospitals a more favorable image to the more vocal segments of the public (while ignoring everyone else [48]) were adopted by numerous hospitals despite protests from many physicians. These protests, however, were moderate (as one might expect), since physicians are subject to the authority of department chairmen and administrators of hospitals. In the future, whenever there was doubt about prognosis, children in general would be treated with maximal effort. Decisions would be based on biological criteria; personal and family feelings about quality of life and about resource limitations would be irrelevant considerations. The American Academy of Pediatrics adopted a position just like this [1, 2] and helped Congress develop legislation which has now effectively made this the law of the land [55]. In a recent report from a major medical center, it appears that a hospital ethics committee has helped to make these constricting reforms part of bureaucratic routine [32].

Incidentally, spokespersons for the American Academy of Pediatrics hold that inappropriate treatment in hopeless situations is far more common than inappropriate non-treatment in situations where there is some hope. One of them told me, "We are clearly more concerned about the evils of selective non-treatment than we are about those of non-selective treatment." This policy fails to protect the larger number of children from futile, abusive treatment, and it fails to protect the budget.

This story is another example of power winning over reason. Who served – and in what ways – to influence this whole process is not clear. But it is clear that from a practical standpoint, the guidelines we drew up are useless at present because a rule mandating treatment in all doubtful situations has been adopted. Space for moral choice is much constricted. Moral communities (including patients, families, nurses, doctors, and others) are enfeebled. As in the first story above, autonomy and creativity are diminished, and it is unclear whether recovery from these losses is possible.

Habermas has described this situation in general, not just for medicine. He notes, "Technocratic consciousness reflects not the sundering of an ethical situation but the repression of 'ethics' as such as a category of life," and adds, "The ideological nucleus of this consciousness is the *elimination of the distinction between the practical and the technical.*" ([27], pp. 112–113) Then technology dominates, and thus introduces confusion of means and ends.

On this point, the reader should take note of the optimism implied behind the pessimism in this chapter. The expectation is that candid presentation of what is may lead eventually to what might better be. Perhaps there is poetic meaning here: just as the seeds of ruin may be found in the flowering of a great movement, so the seeds of better things may be found in ruins. In any case, this is the spirit behind the stories I tell. After one more short story, I will summarize my arguments.

Recently my wife and I met with an attorney to establish living wills and durable power of attorney in the event of incompetence. (At a meeting of the 1985 Norfleet Forum in Memphis, John Fletcher said that people of our ages who neglect to make such provisions potentially are major troublemakers in this era.) We specified that in the event of disagreement between family on one hand and professionals on the other about life support or care in general, the preferences of the family were to be honored; and the family would have to resolve any disagreements occurring among themselves. To this, the attorney responded, "A year ago, I would have opposed that specification. But now, as a result of illness in my own family in the past year, I know better. Although all of us harbor less than noble motives [63], we usually come through in the severest of family crises and make good choices. Besides, at present the medical profession and hospitals in regard to these issues are more concerned about protecting themselves (perhaps for understandable reasons) than caring for patients as conscience dictates."

When I asked the attorney if he thought our wishes would be honored if put to a test, he said, "Avoid testing those things, especially in a court of law." He explained that we should use lawyers and courts rarely.

Even when we do so, we should be reluctant to accept a decision in one instance as a guiding precedent for apparently similar instances in the future. We agreed with him that the reason this approach was best is that there is need to ensure that human uniqueness is protected.

While laypersons commonly organize to exert political action in the interest of defeating disease and death, Lyon reports that families have no organization to which they can turn when confronted with the power of their helpers and that of others whom helpers can recruit ([38], p. 18). It should be noted too that most voluntary health organizations have been created by a combination of laypersons and professionals with common interest in a particular category of disease. These organizations almost invariably seek the *conquest* of their chosen malady. Although humble, heroic acceptance of some conditions may be the most prudent choice [54], these organizations, like the doctors and hospitals we have described, have little patience with that idea.

DISCUSSION

In order to sharpen the foregoing, I will summarize some of the chief characteristics of modern health institutions as these relate to ethics committees. MacIntyre pointed out that physicians in the modern scene necessarily are "bureaucrats," "experts," and "magicians" [41]. They are bureaucrats because hospitals and even office practices are obligate *bureaucracies*. People who work within bureaucracies must develop certain routines for relating to their clients. Physical diseases are treated according to prescribed routines. Patients and families are oriented to play corresponding roles. What is dramatic, frightening, or strange to patients or families is usually routine for personnel in a bureaucracy. There is general reluctance to discuss errors, risks, and alternatives. Dealing with the patient as a person is subordinated to dealing with him or her as a client of a bureaucracy, which to prevent collapse must serve the exacting needs of administrators and technologists.

As a result, doctors and members of the bureaucracy develop one story of the illness which is usually put in *expert*, biomedical terms. The patient has this or that medical condition and is treated accordingly. But there is another story that concerns the illness as experienced by the patient and his or her significant others. Although decisions for care and the care itself should reflect the *integration* of these two stories, such integration is often faulty in modern health care.

Underlying the bureaucracy and the physician as expert is a third condition: the physician as *magician*. I have already described part of that. Frightened, suffering, or sick people want relief, want to live; and it is well known that illness tends to put all of us in the position of dependency. As Katz has pointed out, this dependency is deep and enduring [30]. In illness, we tend to regress to the infant-parent model of relating to others. Since this usually hides the strengths of patients behind the masks of their weaknesses, it tends to buttress decision making within the context of the bureaucracy and within the confines of the physician as biological expert and captain of the ship.

But we don't have to operate that way. If we recognize limits and accept the notion that medical culture, like culture as a whole, may be harmful or helpful, we can be alert to more of the existing possibilities and thus choose more wisely.

From what I have presented, I suggest that using ethics committees for *deciding* patient care is a move toward tyranny – tyranny in the senses earlier discussed and more. Patient and family interests will be subordinated to bureaucratic and biomedical influences, and costs probably will be higher. Alternatively, using ethics committees to *set general guidelines and to nurture moral communities* is a move toward mutual trust and implies in general that committee review and moral community compliance with committee recommendations are optional ([15], pp. 85–95). This second approach could act as a powerful antidote to the traditional paternalism of the medical profession and the progressive tyranny of the technology of medicine, applied necessarily in a bureaucratic setting and sometimes supported wrongly by government. But as I have pointed out, ethics committees are a creation of bureaucracies and are responsible to them, not to patients or families. These committees were designed to deal with some vexing problems that resulted from a 2000-year tradition of paternalism in medicine and a 100-year tradition of great emphasis of biomedical developments and neglect of social and behavioral sciences. At best, these committees can only tinker with the basic problems of neglect of social and behavioral issues. At worst, they may serve to aggravate these problems or cover them up.

In brief, while ethics committees can help to define what is good patient care, they cannot do that which is an integral part of responsible patient care. Such tasks must be the responsibility of moral communities because "the secret of the care of the patient is in caring for the patient" [52].

If we go the route of moral community, I suggest that we can deal with the problems associated with poor prognosis, various reproductive choices, and so on. If we had taken that route in the first place, as Plato proposed for free men, as Codman and Cabot proposed for all people, and as Katz indicated in a recent, deeply penetrating analysis [31], it is unlikely we would be facing some of the most vexing problems we face at present. For example, we would have learned a great deal about coping with limits, with poor prognosis and dying; and we probably would have had the costs of care brought under control.

More specifically, what I propose is that moral communities be given the first chance to solve the problems one by one as they come along. In order that patients and their families may be in charge of deciding the ends of care, they should be fully informed and be the chief decision makers regarding moral aspects of all decisions. This approach will permit patients and families along with their advisors to decide the use of technology more and more on the basis of what is documented to be helpful and less and less on what seems promising or convenient in the opinions of the professionals and the bureaucrats.

This approach involves the acceptance of limits and of some substantial degree of moral pluralism in which there are bound to be disagreements and some errors. We must agree at times that we will disagree – and let it go at that. On this point, Royce gave us some help [58]. But MacIntyre has warned us that in facing disagreements in this age people seem fated to resort to power more than reason [40].

Perhaps the solution to this predicament can be found in further major reforms to complete those begun in the Flexner era. We need a systematic approach to people just as we got a systematic approach to biology. Will faculty seek reforms based on humanitarian interests when they are rewarded most for research in biology? That seems unlikely, because at least temporarily this would involve career and economic sacrifices on their part. Moreover, it would require a shift in "paradigm," the way they think about their work. Kuhn showed that such shifts do not come easily or soon in the scientific community [33]. Will the coming corporation [66], which may dominate the delivery of care, demand reform? That seems possible, but for that an enlightened public and an enlightened corporate leadership will be needed. And they, too, will have to shift their "paradigms" – a change that may not be any easier for them than it is for the scientific community.

I reach four conclusions. First, nurturing moral communities builds strength, fosters healing, and promotes trust. Second, it is not clear whether ethics committees in the present context of caring are capable of nurturing moral communities; they may even harm them. Third, using ethics committees to extend already overextended, unshared decision making enfeebles people, impairs healing, and creates tyranny and distrust. Fourth, the Flexnerian reforms concerning biomedical approaches to the body as a machine appear to be only half of the needed reforms in medicine; and the remaining deficits, many created by technology itself, cannot be patched up by the tinkering of such things as ethics committees. What is needed is a systematic approach to the patient as a person and as a member of a moral community. Only through additional major reforms can we make patient care truly scientific and genuinely humane.

Yale University School of Medicine
New Haven, Connecticut

NOTE

The opinions expressed in this paper are strictly those of the author.

BIBLIOGRAPHY

1. American Academy of Pediatrics: 1984, 'Joint Policy Statement: Principles of Treatment of Disabled Infants', *Pediatrics* **73**, 559.
2. American Academy of Pediatrics: 1984, 'Guidelines for Infant Bioethics Committees'; 1984, also published in *Pediatrics* **74**, 306, 1984.
3. Beeson, P.: 1968, 'Special Book Review', *Yale Journal of Biology and Medicine* **41**, 226–241.
4. Berlin, I.: 1953, *The Hedgehog and the Fox, An Essay on Tolstoy's View of History*, Simon and Schuster, New York.
5. Bloom, S. W.: 1973, *Power and Dissent in the Medical School*, Free Press, New York.
6. Blume, S. W. and Summey, P.: 1978, 'Models of the Doctor-Patient Relationship: A History of the Social System Concept', in E. B. Gallagher (ed.), *The Doctor-Patient Relationship in the Changing Health Scene*, United States Government Printing Office, Washington, D.C., pp. 17–43.
7. Braybrooke, D. and Lindblom, C. E.: 1970, *A Strategy of Decision, Policy Evaluation as a Social Process*, Free Press, New York.
8. Cabot, R. C.: 1915, *Social Service and the Art of Healing*, Moffat, Yard and Co., New York.
9. Calabresi, G. and Bobbitt, P.: 1987, *Tragic Choices*, Norton and Company, New York.

10. Cassel, E. J.: 1982, 'The Nature of Suffering and the Goals of Medicine', *New England Journal of Medicine* **306**, 639–645.
11. Chapman, C. B.: 1984, *Physicians, Law, and Ethics*, New York University Press, New York.
12. Code of Ethics of the American Medical Association, adopted May, 1847, Chapter 1, Article 1, Section 1; or see reference 30, p. 230.
13. Codman, E. A.: 1934, *The Shoulder*, Thomas Todd Company, Boston.
14. Codman, E. A.: 1940, *A Study of Hospital Efficiency*, privately published by the author.
15. Cranford, R. E. and Doudera, A. E. 1984, *Institutional Ethics Committees and Health Care Decision Making*, Health Administration Press, Ann Arbor, Michigan.
16. Cushing, H.: 1940, *The Life of Sir William Osler*, Oxford University Press, New York.
17. Danzon, P. M.: 1985, *Medical Malpractice: Theory, Evidence, and Public Practice*, Harvard University Press, Cambridge, Massachusetts.
18. Duff, R. S. and Hollingshead, A. B.: 1968, *Sickness and Society*, Harper and Row, New York.
19. Duff, R. S. and Campbell, A. G. M.: 1973, 'Moral and Ethical Dilemmas in the Special Care Nursery', *New England Journal of Medicine* **289**, 890–894.
20. Duff, R. S.: 1979, 'Guidelines for Deciding Care of Critically Ill or Dying Patients', *Pediatrics* **64**, 17–23.
21. Duff, R. S.: 1981, "Counselling Families and Deciding Care of Severely Defective Children: A Way of Coping with "Medical Vietnam"', *Pediatrics* **67**, 315–320.
22. Flexner, A.: 1910, *Medical Education in the United States and Canada: A Report to the Carnegie Foundation for the Advancement of Teaching*, Carnegie Foundation, New York.
23. Fox, D. M.: 1986, *Health Policies, Health Politics: The British and American Experience, 1911–1965*, Princeton University Press.
24. Fries, J. F.: 1980, 'Aging, Natural Death, and the Compression of Morbidity', *New England Journal of Medicine* **302**, 938–942.
25. Fromm, E.: 1941, *Escape From Freedom*, Avon Books, New York.
26. Fuller, L.: 1969, 'Two Principles of Human Association', in J. R. Pennock and J. W. Chapman 3 (eds.), *Nomon XI: Voluntary Associations*, Atherton Press, New York, pp. 3–23.
27. Habermas, J.: 1970, *Toward a Rational Society*, Beacon Press, Boston.
28. Hilfiker, D.: 1985, *Healing the Wounds*, Pantheon Books, New York.
29. Henderson, L. J.: 1935, 'The Patient and Physician as a Social System', *New England Journal of Medicine* **212**, 819–823.
30. Kane, R. A.: 1983, 'Social Work as a Health Profession', in D. Mechanic (ed.), *Handbook of Health, Health Care, and the Health Professions*, Free Press, New York, pp. 495–522.
31. Katz, J.: 1984, *The Silent World of Doctor and Patient*, Free Press, New York.
32. Kleigman, R. M., Mahowald, M. B., and Younger, S. J.: 1986, 'In Our Best Interests: Experience and Workings of an Ethics Committee', *Journal of Pediatrics* **108**, 178–188.
33. Kuhn, T. S.: 1962, *The Structure of Scientific Revolutions*, Second Edition, University of Chicago Press, Chicago.

34. Ladd, J.: 1985, 'Philosophy and Medicine', in Eric J. Cassel and Mark Seigler (eds.), *Changing Values in Medicine*, Univ. Publications of American, New York, pp. 203–221.
35. Light, D. W.: 1983, 'Medical and Nursing Education: Surface Behavior and Deep Structure', in D. Mechanic (ed.), *Handbook of Health, Health Care, and the Health Professions*, Free Press, New York, pp. 455–478.
36. Linn, L. S. and DiMatteo, M. R.: 1983, 'Humor and Other Communication Preferences in Physician-Patient Encounters', *Medical Care* **21**, 1223–1231.
37. Ludmerer, K. M.: 1985, *Learning to Heal, The Development of American Medical Education*, Basic Books, New York.
38. Lyon, J.: 1985, *Playing God in the Nursery*, W. W. Norton and Company, New York.
39. MacCallum, M. G.: 1920, 'A Student's Impression of Osler', *Sir William Osler, Memorial Number, The Canadian Medical Association Journal*, pp. 47–50.
40. MacIntyre, A.: 1981, *After Virtue*, University of Notre Dame Press, Notre Dame, Indiana.
41. MacIntyre, A.: 1985, 'Medicine Aimed at the Care of Persons Rather Than What . . .?', in Eric J. Cassell and Mark Seigler (eds.), *Changing Values in Medicine*, Univ. Publications of American, New York, pp. 83–96.
42. Malloch, A.: 'Sir William Osler at Oxford', *Sir William Osler, Memorial Number, The Canadian Medical Association Journal*, pp. 51–65.
43. Marinker, M.: 1975, *The Doctor and His Patient*, Leicester University Press, England.
44. May, W. F.: 1983, *The Physician's Covenant, Images of the Healer in Medical Ethics*, The Westminster Press, Philadelphia.
45. Mayeroff, M.: 1971, *On Caring*, Harper and Row, New York.
46. McCue, J. D.: 1982, 'The Effects of Stress on Physicians and Their Medical Practice', *New England Journal of Medicine* **306**, 458–463.
47. Mishler, E. G.: 1984, *The Discourse of Medicine: Dialectics of Medical Interviews*, Ablex Publishing Corporation, Norwood, New Jersey.
48. Morison, R. S.: 1981, 'Bioethics after Two Decades', *Hastings Center Report* **11**, 8–12.
49. Morris, J.: 1972, 'Three Aspects of the Person in Social Life', in R. Ruddock (ed.), *Six Approaches to the Person*, Routledge and Kegan Paul, London, pp. 70–92.
50. Osler, W.: 1922, *Aequanimitas, With Other Addresses to Medical Students, Nurses and Practitioners of Medicine*, P. Blakiston's Sons and Company, New York.
51. Parsons, T.: 1951, *The Social System*, Free Press, Glencoe, Illinois.
52. Peabody, F. W.: 1927, 'The Care of the Patient', *Journal of the American Medical Association* **88**, 877–882.
53. Plato: 1934, *The Laws*, trans. by W. E. Taylor, J. M. Dent and Sons, London.
54. Preston, R.: 1979, *The Dilemmas of Care*, Elsevier Press, New York.
55. Public Law 98–457, October 9, 1984, *Amendments to Child Abuse Prevention and Treatment Act*.
56. Reiser, S. J.: 1978, *Medicine and the Reign of Technology*, Cambridge University Press, England.
57. Rogers, D.: 1986, 'The Early Years: The Medical World in which Walsh McDermott Trained', *Daedalus* **115**, 1–18.
58. Royce, J.: 1908, *The Philosophy of Loyalty*, Macmillan Company, New York.
59. Scarry, E.: 1985, *The Body in Pain, The Making and the Unmaking of the World*, Oxford University Press, New York.

60. Schaeffer, F. A. and Koop, C. E.: 1979, *Whatever Happened to the Human Race?'* Fleming H. Revell Company, Old Tappan, New Jersey.
61. Schoeman, F.: 1985, 'Parental Discretion and Children's Rights: Background and Implications for Medical Decision-Making', *Journal of Medicine and Philosophy* **10**, 45–61.
62. Seligman, M. E. P.: 1975, *Helplessness*, W. H. Freeman and Company, San Francisco.
63. Shaw, A., Randolph, J. G., and Manard, B.: 1977, 'Ethical Issues in Pediatric Surgery: A National Survey of Pediatricians and Pediatric Surgeons', *Pediatrics* **60**, 588–596.
64. Shklar, J. N.: *Ordinary Vices*, Harvard University Press, Cambridge, Massachusetts.
65. Shorter, E.: 1985, *Bedside Manners, The Troubled History of Doctors and Patients*, Simon and Schuster, New York.
66. Starr, P.: 1982, *The Social Transformation of American Medicine*, Basic Books, New York.
67. Stein, H. F.: 1985, 'Portrait of a Young Physician', *The American Scholar* **54**, 485–499.
68. Stevens, R.: 1971, *American Medicine and the Public Interest*, Yale University Press, New Haven, Connecticut.
69. Stinson, R. and P.: 1983, *The Long Dying of Baby Andrew*, Little, Brown and Company, Boston.
70. Strickland, S. P.: 1972, *Politics, Science, and Dread Disease*, Harvard University Press, Boston.
71. Teel, K.: 1975, 'The Physician's Dilemma, a Doctor's View: What the Law Should Be', *The Baylor Law Review* **27**, 6–9.
72. Thomas, L.: 1974, *The Lives of a Cell*, Bantam Books, New York, pp. 35–42.
73. Viseltear, A. J.: 1984, 'Milton C. Winternitz and the Yale Institute of Human Relations: A Brief Chapter in the History of Social Medicine', *Yale Journal of Biology and Medicine* **57**, 869–889.
74. Welch, W. H.: 1925, *A Great Physician and Medical Humanist, A Review of Harvey Cushing's Life of Sir William Osler*, separately printed by permission of Saturday Review of Literature, November, 1925.
75. Young, J. H.: 1967, *The Medical Messiahs, A Social History of Health Quackery in Twentieth-Century America*, Princeton University Press, Princeton, New Jersey.

NANCY M. P. KING

ETHICS COMMITTEES: TALKING THE CAPTAIN THROUGH TROUBLED WATERS

I. INTRODUCTION

Good navigation is a matter of using a lot of information to make a lot of decisions. You need an abundance of technical knowledge about your ship and crew, the stars, the waters you are in and where you are bound. You need experience to line up the pros and cons of each possible route and destination, as well as the probabilities of successfully executing any maneuver you choose. And you need to be able to map your course according to your needs and goals: where you are going, and how – and how fast – you want to get there. Many of the decisions to be made on a given voyage are easy, straightforward, or indisputable; some are difficult and highly technical; at least a few are likely to be risky or controversial.

Someone who wants to learn good navigation presumably needs to study – and make – all kinds of decisions on all kinds of voyages. But navigation skill cannot of itself instruct the sailor where the ship ought to go – only where it is easy, hard, or impossible to get to. Deciding where to sail is more than just figuring out where we can get to; we need to study more than navigation in order to determine where we should be headed.

Like the role of the navigator on a voyage, the physician's role is intertwined throughout medical decisionmaking and yet is not wholly definitive. The law gives to patients the ultimate right to protect their bodies from even the most well-intentioned interference of physicians ([10]; [13], pp. 114–150). Yet physicians play an essential role in the decision-making of patients by making medical judgments that determine the choices available to the patient and by offering opinion and advice ([25], pp. 76–79; [26], pp. 51–52).

It is easy to conclude, then, that the physician's power to influence decision-making is much more significant than the patient's right to assent to or to dissent from what is offered. Especially when life-and-death choices are at stake, we may be tempted to turn to physicians as

Nancy M. P. King, Larry R. Churchill, and Alan W. Cross (eds.)
The Physician as Captain of the Ship: A Critical Reappraisal, 223–241.

decisionmakers who, as experts, are somehow, we believe, more reliable, more objective, "better" than patients in making crucial choices. We do this in many subtle ways: by encouraging patients to defer, implicitly or explicitly, to the physician's judgment, or by assuming that a patient has done so by submitting to the doctor's care; by allowing physicians to control and limit the information given to patients, upon which they make their decisions; and by finding that patients who are in stressful circumstances or who disagree with their physicians are incompetent to decide for themselves.

It could be argued that this uneasy juggling of decisionmaking influences reflects ambivalence about who really holds the power to decide about treatment choices in medical care [20]. But I don't think that's the right way to understand the situation. When society made its original grant of authority to physicians under medical licensure laws, it was clear that the power delegated to physicians was part of the states' police power both to determine and to provide for the public health and welfare [2, 31]. The state and its people hold the trump card in licensure. At the same time, however, licensing authority is wielded by a selected group of physicians who act as an arm of the state. We delegate authority to experts, and then request that they both control the field and police themselves, using their expertise [16, 30]. Thus, this grant of power and the public expectation that it will be used as we its grantors wish – this *fiduciary* relation – seems not a condemnable and correctable ambivalence but inherent in our best understanding of medicine's mandate. The lively tension in this mandate is well exemplified by medicine's newest decisional entity, the hospital ethics committee.

II. ETHICS COMMITTEES: WHAT SORT OF MODEL?

The reason I have chosen to examine the role of ethics committees in medical decisionmaking is that I am convinced that the phenomenon of ethics committees can help further illuminate the role of medicine in health care decisions. Ethics committees were begun because an inadequacy was perceived that it was believed such committees could remedy. Whether there was then, or is now, agreement about the nature of the inadequacy, or about the effectiveness of the remedy, is another matter. I propose that we will ultimately discover not that medicine's decision-making authority has changed but that professionals, patients, and the public are finally understanding where the authority really was supposed to reside all along.

My claim is that the captain of the ship was never really the captain

[21]. We tend to forget that people usually choose when to see the doctor, and that one who is physically free to refuse to follow the doctor's advice or to change doctors is almost always legally free to do so as well.[1] We have tended to assume until recently – despite much law to the contrary – that when you walk into the doctor's office, or arrive at the emergency room, you have to buy the whole package [29]. Certain crucial junctures in certain kinds of medical scenarios have lately made us more aware, however, that the physician-patient relationship is navigated by choosing among the many successive branchings of a course of treatment. We can, if we want to, pause at times; change course; turn back; or try out the unexplored.

Knowing that the choice is the patient's does not tell us much about ethics committees, though. Ethics committees do not typically play any role in medical decisionmaking when a competent patient is involved. Instead, they address certain cases where the patient is not capable of making a choice, expressing a preference, or participating in the decision. But ethics committees are not involved in all such cases. They usually deal with the hardest cases: the ones that give rise to dramatic uncertainty or conflicting courses of action. Historically, the idea of ethics committees arose in connection with two different kinds of cases: Karen Quinlan-type cases[2] and cases concerning severely handicapped newborns.[3] Sometimes in these cases parents, guardians, or other family members may be involved in making decisions; physicians and other caregivers are always involved; and sometimes ethics committees are added to the crew.

Let me begin not by considering hospital ethics committees but by comparing two other common varieties of hospital committees: what is usually called the morbidity and mortality conference, and the Institutional Review Board (IRB). Ethics committees bear some similarity to both of these, and could be considered as falling somewhere between them as regards their powers, duties, and goals.

Morbidity and mortality conferences are medical staff peer review activities whose stated purpose is to review retrospectively adverse medical outcomes in order to oversee the quality of care rendered by the institution. Such conferences might be analogized to the review of a completed voyage – Columbus' report to Ferdinand and Isabella. They are composed of members of the medical staff and their function is usually prescribed in the staff bylaws. For staff physicians whose treatment decisions and actions come before them, morbidity and mortality conferences serve two basic functions.

First, they provide feedback and guidance regarding the standards of the institution and of proper professional practice, while permitting the institution to stay informed about the care it is delivering. This pedagogical and monitoring function helps to ensure the delivery of good medical care. The morbidity and mortality conference is not perceived as intruding upon the physician's exercise of medical judgment. Instead, it acts appropriately as part of the professional and institutional effort to educate and correct that judgment continually according to the standards of the medical profession itself.

Second, the morbidity and mortality conference has the related but less easily understood function of ratifying the individual physician's experiences and choices. This aspect of such a committee's role has been accurately and vividly described by a prominent medical sociologist as to "forgive and remember" [7]. The experience of appearing before a morbidity and mortality conference provides vital professional support and solidarity in the face of the uncertainties of practice. Thus, this type of committee reinforces the physician's role in difficult, life-and-death cases, rather than altering it.

IRBs were conceived as representing a rather different committee model. First instituted to protect the rights and welfare of human subjects of federally funded research ([13], pp. 208–221; [14, 23, 28]), IRBs pass prospective judgment on proposed research, evaluating both its scientific merit and its plans to safeguard subjects from physical injury, from psychic outrage, and from any failures to consider and treat them as full partners in research. The IRB's review of research is like a pre-flight check, the last inspection of ship and crew before weighing anchor and setting sail.

The composition and function of IRBs are prescribed by federal regulation [26]. Their membership is generally predominantly physicians and scientists, but there is also some lay representation. The two purposes of IRBs – to protect human subjects from harm and to ensure that they are treated with respect – reflect two functions.

First is the intraprofessional monitoring of research according to the standards of good practice ([13], pp. 208–209; [26], sec. 46.102(b) (1)). When an IRB determines whether a research protocol has scientific merit – that is, whether it is designed to produce useful knowledge and benefit at the cost of a not unreasonable degree of risk to subjects – its function is analogous to that of a morbidity and mortality conference, except that the IRB can actually prohibit a researcher from taking a particular course of action.

In addition, however, the IRB has a second role: to ensure that the researcher obtains the human subject's informed consent to the procedure as a means of recognizing and respecting the subject's humanity, autonomy, and partnership in the research process ([13], pp. 208–209; [26], sec. 46.102(b), (2) and (3)). This very different function is not incompatible with the professional goals of good science, but it does not spring intrinsically or necessarily from the profession itself. Instead, it is a value externally imposed by society on the scientific enterprise: it shapes and changes what researchers do. This respect for human subjects is a concern not of science as science, but of science as an enterprise within the human community. (In fact, it has been enforced by courts, by governmental regulation, and by other such forums and tribunals more than by scientists themselves; [13], pp. 151–232.)

Thus, it is fair to say that IRBs were intended to – and do – change the very idea of what scientists are supposed to do, whereas morbidity and mortality conferences are intended to correct what doctors do but not to change what they are supposed to do. I shall argue here that hospital ethics committees fit somewhere between these two committee models.

Ethics committees – which are organized in many different forms and have many different names as well ([24], pp. 440–441; [28]) – are difficult to generalize about, but seem to fall nicely between our two other committee models, at least in a temporal sense. Ethics committees are usually described as having the following functions: (1) making policy and establishing information and education resources; (2) reviewing decisions retrospectively; and (3) consulting about cases that are currently in the midst of decision ([1], p. 5; [3, 4]; [14], p. 2688; [22], p. 14894; [24], pp. 440–441, 451). The first, policymaking function is analogous to the role of the IRB; retrospective review is like what the morbidity and mortality committee does; and, finally, concurrent consultation is neither before nor after the voyage but takes place right there on the deck, in the midst of troubled waters.

Perhaps the most potentially far-reaching role for ethics committees is policymaking. Committees are to make institutional policy where possible – that is, to decide which cases should come before the committee as well as to establish guidelines for cases where treatment, or non-treatment, is clearly required. Policymaking affects physicians' decision-making long before cases come to the committee. For example, an infant care review committee might develop a policy stating that Down's syndrome alone cannot provide a basis for withholding treatment for other health problems, or that anencephalic babies need not be maintained on respiratory support if the parents concur. Of course, policies will only be possible for relatively clear-cut and easy "prognosis"-type cases, like the examples just given. Many committee consultations may arise within the fuzzy edges of such policy – for example, how much brain mass may an anencephalic child have and still be labeled anencephalic? Is that a strictly medical determination, even if it amounts to determining when a child need not be maintained? What about when the parents seek to maintain their anencephalic child, who is, functionally speaking, in a vegetative state or even brain-dead? Many such policy interpretation cases will continue to come before even a committee that has set policy. And I would argue that dealing with particular cases is more central to the nature of an ethics committee than even this vital policy-making role.

Thus follows the committee's second function, which is to ensure that health care providers have all the information and support they need to make reasonable medical judgments about severely ill patients. This is a continuing education function on a small scale, which can be accomplished simply by ensuring that house staff and attending physicians have regular opportunities to share updated knowledge about relevant conditions, treatments, and outcomes.

The third function of an ethics committee is to make certain that patients' families are afforded the opportunity to understand the medical facts, and to make available to them and to their caregivers all information relevant to the care, treatment, and financial and social support of handicapped and disabled individuals in the community. This cataloguing of resources is vital – not as a means of facilitating quality-of-life decisions but simply as providing information necessary to the decision-maker's understanding of the outcome of a decision to treat. For example, parents of a child with myelomeningocele need to know something about the number of operations their child will face in the first years of life, as those surgeries are consequences of the decision to treat. Similarly, the family of a severely brain-injured patient needs to know whether the patient will have access to rehabilitation programs that could restore some cognitive capacity, since the availability of such programs may change the possible outcome of a decision not to terminate life-supporting care.

Finally, ethics committees advise patients, families, and caregivers by examining cases before them and determining whether they agree with the proffered decisions. This advice is based on the decisional guidelines each committee sets for itself. Many of the cases that come before ethics committees are retrospective reviews. This may in part be because it appears difficult, cumbersome, and overly dramatic to convene the committee for a special meeting; but in most of the cases that are reviewed retrospectively there was no disagreement between physicians and family that would have necessitated a prospective review. Infant care review committees review many cases of severe birth injury or congenital malformation or extreme prematurity, where the physician was certain of the prognosis and sought and obtained familial concurrence to discontinue respiratory and other support, or to institute a "no-code" order or an order not to treat infections and other complications that might develop while leaving the respirator in place. Of course the most difficult situations are most likely to arise as concurrent reviews. Then the crucial question becomes one of the committee's role and authority when it acts as advisor.

Two physicians, Ronald Cranford and Norman Fost, have done much writing abut ethics committees that simply and clearly illustrates the variety of models that these committees might adopt for concurrent review [14]. The most useful schema they offer is a basic box diagram, with one axis representing the consultation requirement and the other the compliance requirement. In every institution that establishes a committee to play some concurrent role in some decisions, two principal questions must be addressed: (1) Is consulting the committee mandatory or optional? (2) Is following the committee's recommendation mandatory or optional? Plugging these into the box diagram produces four models: mandatory-mandatory, optional-optional, mandatory-optional, and optional-mandatory ([14], p. 2691). Not unexpectedly, the second and third of these models – where following the committee's recommendation is optional – are by far the most prevalent.

Indeed, the committee is not supposed to make the decisions; mandatory compliance committee models are legally flawed. The committee can have no final decisional authority. It has been well understood in law since at least the early part of this century that competent adult patients are the ultimate authority for their own health care decisions ([10]; [13], pp. 114–150; [21]). If the patient cannot choose, or is not an adult, a legal guardian may decide; but all decisions by others are potentially subject to legal challenge and ratification. Thus, one who disagrees with a committee's recommendation is always free to bring the case into the courts, seeking a judicial order to initiate or cease the disputed treatment ([1], p. 7; [21]). Indeed, the committee itself might be viewed as having a duty to bring some still-disputed cases to the attention of a court for an enforceable, final decision.[4]

So committees don't decide, and their deliberations represent only an interim step between decisions by physicians and family and decisions by the courts. What then are they good for? Why do physicians and families seek their counsel?

It seems that the physicians who feel comfortable about consulting the committee often do so to validate their decisions, especially in cases where there is no potential disagreement with the family. Most such cases are relatively "easy" ones, and the decisions reached are often relatively "conservative;" the greatest difficulty involved is the emotional difficulty of dealing with the family. In such cases physicians seem to turn to the committee for reasons similar to those operative in morbidity and mortality conferences. Here physicians' ethical judgment is ratified, their ability to deal with important moral and societal concerns is reinforced, and they can reassure themselves and the committee that things are going along as they should.

Sometimes these cases may seem so clear that they almost need not have come before a committee. The physician's perspective may be that though it does seem clear, there is still reason to be sure the committee thinks so too, as it is difficult to differentiate the clearcut cases from the cases that are genuinely disputable. Besides, small differences in developing medical techology have the potential to affect the certainty of medical judgment, and therefore the appropriateness of decisions regarding treatment. The committee benefits from learning what is currently viewed as medically feasible and from being alerted to what might change that. In short, science and society have not yet reached the point where it makes sense for an ethics committee to consider it unnecessary to hear any case physicians or family wish to bring before it.

Physicians seek the committee's reinforcement of their decisions for other reasons too, which also range beyond what the committee is "for" in the strictest sense. In bringing together its wide range of disciplines, skills, and outlooks, the ethics committee may be the easiest place to go in the hospital for informal advice on legal issues, suggestions as to how to raise difficult issues with families or with other physicians, and review of the ramifications of different courses of action. An important and unexamined role for the ethics committee lies in its acting as a forum for the exploration of differences in moral judgment among members of a team of caregivers, so that decisions arrived at may be shared, understood, and ultimately supported by all parties involved.

The committee may (1) suggest consultation with different experts or community groups; (2) help walk through the physician's proposed decision and its consequences; and (3) simply serve, collectively, as another listener who may ask questions as yet unasked and think of things not yet thought of. Fundamentally important too is that merely formulating and presenting the case before the committee – this public act of sharing – changes it for the physician, and may itself bring new, unexamined aspects to light. The need to spell out the things that shared expertise has until now kept in medical shorthand, combined with the candor possible before non-physicians who are not the patient's family but one's peers, gives the committee's process its own informal yet ritual gravity. One can't be sure precisely how the experience of consulting an ethics committee will affect particular decisions, but I would speculate that they are affected simply by the expectation that such a consultation is worth doing.

Parents and families of patients may be presumed to want some different kinds of help from the committee. Perhaps the parents have sought the committee's consulation because they have been told that they don't have to decide about their child – the committee is supposed to do it. Or the committee's agreement with the physicians' proposed course of action may provide important support for one parent in the attempt to untangle a disagreement between the parents regarding their child's care.

In most cases it seems likely that the information, and especially the resource and community support network, that the committee can provide will be of greatest significance to parents and family. After all, the committee will not appear to family as the collection of colleagues and peers that physicians may find there; instead, it is likely to seem a group of strangers with an ill-defined responsibility for passing judgment on parental choice. Not surprisingly, as a committee becomes more experienced in prospective consultations with families of patients, it is likely to become better able to address families' needs and concerns.

This review of the functions of ethics committees is useful, of course, but it does not really get at the heart of what these committees are about. Simply put, what ethics committees do is step into medical decision-making when the patient is in no position to make or communicate a decision, when the choices are often between life and death, and when neither the correct decisionmaker nor the correct decision is at all clear.

Recall that, as I have already said, the choices about medical treatment belong to the patient when the patient can make them.[5] But remember also that the physician's role as healer and expert advisor is firmly established, both by the sanction of society and by the individual patient's need for a skilled and trusted agent acting in his best interest. Why shouldn't we view the physician as the correct decisionmaker when the patient cannot decide?

I believe that the reason we should not – and do not – give physicians exclusive power in this regard is that medical judgment does not have all the answers we need. The authority that society grants to physicians is based on their special expertise, but medical expertise has limits. Every decision – no matter how "scientific" – contains elements of the decisionmaker's values, whether personal or professional. The very fact that law and society recognize the patient as the ultimate decisionmaker demonstrates that the members of society do not necessarily accord the professional values of the physician the same deference that is given to the physician's scientific knowledge and skill.

Indeed, in the hardest choices it is clear both that the role of values is significant and that the values involved – and the decisions reached – may significantly differ. Simply put, there is no such thing as a purely medical decision. There are nonmedical aspects to every choice, and no medical decision ought to be made without taking account of them. But it then remains to identify the decisionmaker who is charged with considering the values in the choice. Perhaps there are still many instances when we can, and do, give that role to the physician. But we have come to recognize, now more than ever, that making these value-laden choices on behalf of everyone else is not something doctors are supposed to do. When patients can decide for themselves, they place their own weight on these values for themselves. Similarly, when communities decide, as matters of public policy or law or the organization of health care delivery, to consider non-medical values in a particular way, physicians and patients, as community members, may be properly bound by those value choices. Yet the community's power is ultimately limited; we can reach only so far into the privacy of each other's lives. The value-laden choices we make about our lives and our bodies are the most private of all.[6]

I believe that the current posture that encourages the establishment of hospital ethics committees stems from the growing knowledge that no special training equips any particular professions or professionals to consider the many personal value questions necessarily involved in difficult medical treatment decisions, such as decisions to withhold or withdraw lifesaving treatment from persons unable to make their own choices. Therefore, the boundaries of the unique authority conferred on the ethics committee are drawn by moral uncertainty. Within the boundaries is contained a societal agreement to disagree: we are certain only that we are not certain about how to handle these hardest cases. That origin tells us that ethics committees are not tribunals. They do not decide things. We didn't create them to give us the right answers, but rather to help us acknowledge that we don't always have the right answers.

III. ETHICS COMMITTEES AS MORAL COMMUNITIES

The desire for right answers is seductive and may be overwhelming. In the realm of the hard cases we are discussing here, *rightness*, because it cannot be supplied by science, is often expressed instead in terms of *rights* – the right to life, the right to practice one's profession, the right to a peaceful, "natural," or dignified death. But the realm of uncertainty I have outlined here seems incompatible with this invocation of rights language. How can we assert that no unique choice is the correct one, if there are rights to be protected? Rights must always either win or lose out. The question is whether rights as we typically understand them have meaning and usefulness in the hard cases faced by ethics committees. I believe that they do not, but that ethics committees are needed because they provide some important and useful substitutes for "rights" in these cases.

My analysis here is heavily indebted to Martin Golding's work in progress on moral community, rights theory, and dispute resolution [15]. He describes three basic types of dispute settlement mechanisms. *Adjudication*, an adversarial, rights-oriented system, is the province of courts and arbitration schemes. In contrast are *conciliation* – best typified by mediation and other bargaining schemes – and *psychic transformation*, the characteristic form for counseling and analysis. These latter two dispute-settlement mechanisms are not rights-oriented. Instead, it might be said very broadly that, whereas adjudication attempts to sort out the right answer, conciliation and psychic transformation focus on the transforming quality of the decisionmaking process itself.

Golding views each of these three mechanisms as mini-communities (temporally as well as in size), and argues that the appropriateness of their use depends, in each case, on their fit with the parties and circumstances that will constitute the temporary community. For example, because adjudication is by its nature rational, factual, objective, and distanced, it is often disruptive and destructive in intrafamilial disputes. It is relatively uncontroversial to say that going to court is likewise not a good way to settle disagreements between families and physicians in the cases at hand. It *is* controversial, however, to take the next step and say that therefore, the notion of rights does not apply. Some have made this argument by treating the family unit as a unique entity, a seamless web that ought not to be invaded even in the name of the ill, voiceless member's "rights" [9]; but I find that argument unsettling and unsatisfying. More satisfying, it seems to me, is Golding's view that a different, non-adjudicatory community may provide a fully adequate alternative.

In my view, Golding's two non-adjudicatory models – conciliation and psychic transformation – offer substitutes for rights that satisfy our needs for moral identity and moral activity in much the same way as rights do. Mediation and counseling both offer all parties a voice, a moral language, a scrupulous and respectful ear, and acknowledgment of the reality of the problem and the effort entailed in its resolution. This recognition of the involved parties is real, and substantial, and satisfying. Moreover, it adds an important dimension: the recognition that the procedure of making these decisions is a challenging – and ultimately transforming – work.

Golding's two non-adjudicatory communities specifically lay our conventional notion of rights aside in favor of the transforming recognition of moral agency afforded by their very different decisionmaking procedures. The work of ethics committees is somewhat different. Ethics committees do not make decisions. This difference does not weaken the significance of ethics committees, however; instead, it provides us with the opportunity to probe a little deeper into our assumptions about rights in medical decisionmaking.

We are accustomed to thinking of rights as trumps: the cards that win the game [11]. We tend to assume that when rights are involved, all that is required is that somebody sort it all out correctly and find the answer – and that's the end of the matter. In situations of difficult medical decisionmaking it readily becomes clear that rights understood in this way give rise not to answers but to impasse, and moreover that any declaration of a winner is only the beginning. It matters who decides, and how a decision is reached, because whatever choices and actions follow are part of the continuing relation of patient, physician, and family, not just a judgment executed by an invisible hand.

In the setting of medical decisionmaking, then, rights are properly understood not as trumps but as ideas that help define the moral identities of those facing decision. Those who are deciding – physicians and family – and those who are being decided about – patients unable to make or express their own views – are bound together in the decision, and the moral identities of all of them must be recognized and honored. I believe that only an organized and selfconscious forum, like that provided by an ethics committee, can afford that recognition. The ethics committee is a moral community – temporary, confidential, but also public in a limited sense, affording an open and respectful hearing to all concerned and returning an acknowledgment of the meaning, validity, and value of the decisionmaking process.

The committee is a forum whose function it is acknowledge the difficulty of the decisions it considers and to offer support – in a sense, to celebrate the process of decision and its transforming effect on the decisionmakers. This is like the "forgive and remember" purpose of peer review, but it is also more than that. Kathryn Hunter has described the similar institution of critical care review grand rounds as a kind of Greek chorus of sympathetic commentary on these hard cases and memorialization of the patients involved, who might otherwise be too easily forgotten [17]. In addition, the chorus itself, by its participation and commentary, must also be transformed.

Uncertainty – whether it be moral or factual – is difficult to face when action is necessary because it bears on our own sense of fitness as moral agents. My view of the ethics committee's role as celebrator of decisions made under uncertainty could be challenged because it appears to acknowledge neither *rights* nor *right*. Yet my argument is that the ethics committee does acknowledge both, in the *rite* by which it recognizes the decision and its subject and makers.

The ethics committee begins with the certainty that no absolute certainty exists – thus reinforcing the modern perception that the physician's expertise can accord him no more authority than *medical* skill and knowledge provide. But the committee does not interpose a special *ethical* expertise in place of the old, incorrect assumption that the physician knew best. Instead it proposes that our own moral sensibilities – those of physician and non-physician alike – can be exercised responsibly *when certainty eludes us*. And finally it embodies the hope that by dint of careful and continuing public examination, those moral sensibilities can develop, not into absolutes, but into a practiced and conscientious moral reasoning.

Education, advice, policymaking, retrospective review, information coordination – all are important functions of ethics committees, but do not themselves justify an ethics committee's existence as a unique entity. All of those things can be done (and many *are* done) by other hospital offices, groups, and individuals.

That committtees may uncover faulty, insensitive, or unprofessional decisionmaking cannot alone justify them either. Any number of different kinds of review, from peer processes to the courts, can through increasing vigilance decrease the number of "bad" decisions. If we were all agreed that too many "bad" decisions were being made about care in hospitals, then instituting an ethics committee would be one way – but only one of many – to detect and prevent such decisions. Another way might be to require that all such decisions be reviewed by a court. Another would be to give an in-hospital administrative tribunal the power to decide. Still another would be to link demonstration of decisionmaking skills to staff privileges or even to licensure. Thus it can readily be seen that if "better" decisions are the goal, committees are only one means to them. For committees this goal, though important, is still incidental. And since committees as they exist now have no legal authority to make decisions, if "good" decisions are the point, committees are a comparatively poor way of getting to the point.

If we were all agreed that bad decisions were being made on behalf of persons not able to decide for themselves – that is, if we were agreed about *right* and *rights* – we would have moral and legal obligations to intervene. What keeps the ethics committee from being a *deciding* body is the fact that it derives its mandate from the cases about which we cannot agree. The essence of the committee's role is uncertainty; its scope is determined by moral dilemma. It operates where we have no clearly right answers; and therefore its means of operation is a *rite*, confirming that responsibility for a particular answer reposes with the family and physician, its moral heirs.

IV. CONCLUSION

Do committees then have moral authority, in this dim and difficult undecided realm of uncertainty? Yes, I believe that they do, but it is moral authority of a unique and particular sort. Committees are not intended to be moral experts, composed of persons more able or fitted to address moral questions than are patients, their families, and caregivers. So a committee's authority does not lie in its members' superior abilities.

Rather, the ethics committee's moral authority is provided simply by acceptance of its value as a forum. American society recognizes the legal authority of the court system because we understand the courts as products of our social system; we embrace them as such and therefore feel responsible, as a society, for ensuring their working well. Belief in the jury system and some degree of understanding of the workings of the criminal law are central to popular political culture; and we are keenly aware that our continuing faith in the courts depends on their demonstration of openness, fairness, and faithfulness to the principles they purport to uphold. Though most people's involvement in the judicial system is minimal, as a people we have a strong and vital stake in it.

Similarly, an ethics committee derives its moral authority from its own conscientious attempt to define its role, to carry out its functions faithfully, and to engage in self-examination. Unlike a jury, the committee does *not* apply community values in reaching conclusions about the behavior of members of the community. Instead the committee represents the community's agreement to disagree, by identifying the medical decisions that entail more dilemmas so important and so problematic as to require, first, the special consideration of a special body and, second, the community's support for any decision conscientiously made. The committee's special consideration and support supply for the decision the same kind of validity that is supplied by a court's determination of rights.

This is the heart of a committee's authority: the identification of those problems where there is really no finally, clearly right answer and above all no easy answer. Where all choices are difficult, painful, and demanding, there is an important and unique role for a body like an ethics committee: to listen and reflect, to acknowledge and affirm the deliberations of family and caregivers. In the course of this process – which is a ritual recognition of the *solemnity* of the decisions brought before it –

the ethics committee will, inevitably, educate and inform, uncover decisions poorly made, and give advice that will facilitate more conscientious decisionmaking. But these are, in an important sense, extra bonuses, albeit of great value.

I do not wish to be misunderstood in my main point; it is crucial that we understand the ethics committee as doing three dissimilar but inseparable things. First, committees define – or reflect the community's definition of – the class of medical decisions for which there is no right answer and in which, therefore, decisions are properly left with families and caregivers. Second, committees open up the deliberative process, providing matter and opportunity for retelling, rethinking, and reflection about both particular decisions within these boundaries and the realm of decisions so regarded. And, third, committees have a function that is both ritual and supportive, affording public recognition of the solemnity – and, if you will, nobility – of choices that are tragic and conscientiously undertaken, and of the corresponding moral agency of the choosers.

What I describe is plainly an ideal of sorts. Without doubt there are ethics committees that have been "captured" by the bureaucratic concerns of their institutions or by political fears about the correct line to take. Certainly even the most conscientious of committees finds itself often resented by caregivers, family, or both, calling into question in another way the validity of its community-representative role. But my own direct experience of ethics committees leads me to conclude that much of what I would claim for them is at least sometimes true.

I spent six months serving on the most common variety of ethics committee, an infant care review committee. In that brief time I saw a wide range of physician responses to the committee, from refusal to answer its questions, to expressed anger, to misconceptions about its role, to genuine seeking and sharing with it. I saw the committee itself discuss a wide range of cases compassionately and with perspicacity, and I saw it moved by those cases and by the moral and emotional strength and skill displayed by parents and physicians in their decisionmaking. In short, I saw the committee work precisely as it was meant to most of the time.

In particular, I saw that those physicians who came to the committee regularly benefited in two ways. Expectably, they gained insight that they were increasingly able to apply in their decisions. But they also gained something distinctive from *every* presentation they made: that extra insight and support from public presentation, discussion, and solemnization of their responsibilities as professionals, advisors, and moral actors facing uncertainty.

To our opening question, then: Do ethics committees change what doctors do? Yes, in two ways. First, they underscore that doctors should not do what they are not supposed to do: make the final decisions that belong rightfully to others in a shared community [12, 20, 21]. Committees help illuminate the entanglement of science and values and preclude naive attribution of a simplistic and unreflective authority to medical judgment. Second, they validate the physician's exercise of moral reasoning, in those very instances when the physician's judgment will not be decisive. This is not a contradiction. Physicians are moral agents like everyone else, but it is perhaps a great change in their image to declare – as our society has done in implementing ethics committees – that they are not the *sole* moral agents to act in medicine. Ethics committees reinforce our realization that moral agency requires some training in and exercise of knowledge and judgment. We do not need moral *experts*; but neither will we accept moral *pronouncements*.

For a long time medicine was our altar of the infinitely possible. Now we have come to recognize the distortion in that view, but we are still far from accepting that certainty is rarely available. The ethics committee's role in reforming our collective sense of appropriate expectations from medicine and appropriate responsibilities in medicine is just one part of this effort. We will not discover the right answers through ethics committees. Instead, we will come to appreciate a kind of rightness that results from the willingness to reflect and to question, to disagree and to learn together.

School of Medicine
University of North Carolina
Chapel Hill, North Carolina

NOTES

[1] Most court-ordered or otherwise enforced treatment of arguably competent patients – e.g., that of Elizabeth Bouvia [6, 19] or Dax Cowart [8] – has been directed toward those with little or no physical freedom. Courts have on occasion ordered treatment of compe-

tent, free, and mobile pregnant women in the interest of their unborn children, but this questionable power has not been exercised to seize and detain women not voluntarily within a health care institution at the time of the order. See [5].

[2] The Quinlan case [27] was the first judicial call for ethics committees. The court in that case relied, *inter alia*, on a call by a physician for the diffusion of responsibility in decisions to withhold or withdraw lifesaving treatment from incompetent patients. See [12]. The paradigmatic Quinlan decision dealt with the withdrawal of artificial respiratory support from a young woman in a permanent vegetative state. Though she was expected to die when the respirator was withdrawn, Karen Quinlan lived 10 years in a nursing home until her death from pneumonia.

[3] After a much-publicized case where lifesaving surgery was withheld, with court approval, from an infant with Down's Syndrome, the Federal government promulgated regulations about such decisions that included guidelines for the establishment of Infant Care Review Committees [22]. Such committees were not required by the regulations, but a great many hospitals put them in place nonetheless. See [1, 3, 4, 18].

[4] The Federal regulations ([22], p. 14896) state: "[T]he ICRC should counsel that the hospital board or appropriate official immediately refer the matter to an appropriate court. . . ."

[5] There are two circumstances confounding this otherwise simple assertion, which I will not discuss in this paper. The first is the possibility that a competent person's choices may not be available to the decisionmaker, as in conditions of severe physical handicap, e.g., locked-in syndrome. The second is the problem of determining the competence of a choice expressed by an arguably incompetent patient.

[6] One of the practical reasons for ensuring that such decisions belong ultimately to patients themselves appears not infrequently in medical settings. This is the risk that when more than one physician is involved in the care of the patient, the physicians may disagree about these value choices. Reposing the decision with the patient, though it does not prevent disagreement among family and friends purporting to speak for the patient, at least focuses authority in a way that is likely to eliminate other disagreements.

BIBLIOGRAPHY

1. Adjunct Legal Task Force on Biomedical Ethics: 1985, *Report, Legal Issues and Guidance for Hospital Biomedical Ethics Committees*, American Hospital Association, Chicago.
2. Aitchison v. State, 204 Md. 538, 105 A.2d 495 (Md. Ct. App. 1959).
3. American Academy of Pediatrics: 1984, *Guidelines for Infant Bioethics Committees*, AAP, Evanston, IL.
4. American Hospital Association: 1984, *Guidelines, Hospital Committees on Biomedical Ethics*, AHA, Chicago.
5. Annas, G.: 1982, 'Forced Cesareans: The Most Unkindest Cut of All', *Hastings Center Report* **12**, 16–17.
6. Annas, G.: 1984, 'The Case of Elizabeth Bouvia', *Hastings Center Report* **14**, 20–21.
7. Bosk, C.: 1979, *Forgive and Remember*, University of Chicago Press.
8. Burt, R.: 1979, *Taking Care of Strangers*, The Free Press, New York.
9. Burt, R.: 1977, 'The Limits of Law in Regulating Health Care Decisions', *Hastings Center Report* **7**, 29–32.

10. Capron, A.: 1978, Right to Refuse Medical Care', in W. Reich (ed.), *Encyclopedia of Bioethics*, Vol. 4, The Free Press, New York, pp. 1498–1507.
11. Churchill, L. and Siman, J.: 1982, 'Abortion and the Rhetoric of Individual Rights', *Hastings Center Report* **12**, 9–12.
12. Duff, R.: 1988, 'Unshared and Shared Decision Making: Reflections on Helplessness and Healing', in this volume, pp. 191–221.
13. Faden, R. and Beauchamp, T. with King, N.: 1986, *A History and Theory of Informed Consent*, Oxford University Press, New York.
14. Fost, N. and Cranford, R.: 1985, 'Hospital Ethics Committees', *Journal of the American Medical Association* **253**, 2687–2692.
15. Golding, M.: *Moral Communities* (manuscript in preparation).
16. Gross, S.: 1984, *Of Foxes and Henhouses: Licensing and the Health Professions*, Quorum Books, Westport, Ct.
17. Hunter, K.: 1985, 'Limiting Treatment in a Social Vacuum', *Archives of Internal Medicine* **145**, 716–719.
18. Judicial Council of the American Medical Association: 1985, 'Guidelines for Ethics Committees in Health Care Institutions', *Journal of the American Medical Association* **253**, 2698–2699.
19. Kane, F.: 1985, 'Keeping Elizabeth Bouvia Alive for the Public Good', *Hastings Center Report* **15**, 5–8.
20. Katz, J.: 1984, *The Silent World of Doctor and Patient*, MacMillan, New York.
21. King, N.: 1987, 'Federal and State Regulation of Neonatal Decision-Making', in R. Macmillan (ed.), *Euthanasia and the Newborn: Conflicts Regarding Saving Lives*, D. Reidel, Dordrecht, Holland, pp. 89–115.
22. Model Guidelines for Health Care Providers to Establish Infant Care Review Committees: 1985, *Federal Register* **50**, 14893–14901.
23. National Commission for the Protection of Human Subjects of Biomedical and Behavioral Research: 1978, *Report and Recommendations: Institutional Review Boards*, United States Government Printing Office, Washington, DC.
24. President's Commission for the Study of Ethical Problems in Medicine and Biomedical and Behavioral Research: 1983, *Deciding to Forego Life-Sustaining Treatment*, United States Government Printing Office, Washington, DC.
25. President's Commission for the Study of Ethical Problems in Medicine and Biomedical and Behavioral Research: 1982, *Making Health Care Decisions*, United States Government Printing Office, Washington, DC.
26. 'Protection of Human Subjects', 1985, *Code of Federal Regulations* 45, Part 46, United States Government Printing Office, Washington, DC.
27. In re Quinlan, 70 N.J. 10, 355 A.2d 647 (1976).
28. Rosner, F.: 1985, 'Hospital Medical Ethics Committees: A Review of Their Development', *Journal of the American Medical Association* **253**, 2693–2697.
29. Shipley, W. E.: 1967, 'Liability of Physician or Surgeon for Extending Operation or Treatment Beyond that Expressly Authorized', *American Law Reports 2d* **56**, 695–706.
30. Starr, P.: 1982, *The Social Transformation of American Medicine*, Basic Books, New York.
31. State ex rel. Schneider v. Ligett, 223 Kan. 610, 376 P.2d 221 (1978).

NANCY M. P. KING

AFTERWORD

Pressures on the authority of medicine – what these essays have called the "captaincy" – come from two distinct directions: from the larger realm, in which society delegates to physicians the power to provide for the health and welfare of its citizens, and from the smaller, more immediate realm wherein each patient and physician negotiate about the exercise of the power to make decisions about treatment. In the larger realm – as we have seen, for example, in the essays by Russell Maulitz, Joan Lynaugh, Ernest Kraybill, Robert Cook-Deegan, and Stuart Spicker – the physician's skill and knowledge is challenged as a basis for professional power on several fronts: by other professionals (both new and old competitors); by institutional managers; and by government itself. One source of these challenges is disagreement about what skill and knowledge is most needed so that public welfare goals may be accomplished. Another source, one which is growing in significance, is confusion and disagreement about the goals themselves: how shall we balance the goal of health against other pressing welfare and non-welfare goals, such as housing, education, freedom and diversity, even financial stability?

Between patient and physician, on the other hand, it emerges ever more clearly – as discussed, for example, by David Barnard, Raymond Duff, Judith Areen, and Peter Morris – that patient-centered medical practice means more than Hippocratic devotion to the patient's health; it must include respect for the patient's decisional autonomy as well. The pressures on physicians in this realm constitute a different kind of challenge to professional power: not to the physician's skill and knowledge but to the authority for the use of that skill and knowledge. This challenge may be the more fundamental; it identifies as false some common assumptions and practices regarding authority in the physician-patient relationship. Nonetheless, it may be an easier challenge than the first to meet. Once the goals of the physician-patient encounter are determined by the physician and patient together to their satisfaction, the physician's service ethic can address the challenge of fulfilling those goals by applying professional knowledge and skill.

Nancy M. P. King, Larry R. Churchill, and Alan W. Cross (eds.)
The Physician as Captain of the Ship: A Critical Reappraisal, 243–246.

One outcome of our examination of the authority of medicine is the realization, touched on by H. Tristram Engelhardt, Wendy Mariner, and Nancy King, that these two distinct types of challenges to it – from the larger society and from individual patients – are related in a necessary way. First, there are important connections between the authority of skill and knowledge and the social and legal power of action and control, of "captaincy." Second, society's grant to the medical profession of the authority to provide for the welfare of its citizens parallels the patient's grant of authorization to the physician in an individual encounter. In both realms, power originates not with the profession but with the recipients of professional service; yet much of that power is transferred to the profession, by virtue of its knowledge and skill, so that its service goals might be best accomplished.

But can recognition of these relationships between the societal and interpersonal realms, between being *in* authority and being *an* authority, between medicine and the society it serves, be of any help in charting medicine's future course? The symposium from which this volume is drawn concluded its metaphorical examination of the authority of medicine with a panel discussion entitled "Health and the Community." The focus of this discussion was the recognition of a most salient relationship: Communities, composed as they are of persons who are members of society and who are also, necessarily, patients and patients' families, represent a point of intersection between the societal and interpersonal realms.

The perspectives of the community in an examination like ours may often be unfamiliar and are sometimes quite new. Health and health care form but one part – an integral yet far from all-encompassing part – of the life of the community and of individuals' lives. Patients and families view health care as something to be fitted in amid the rest of their day-to-day concerns, a viewpoint that is sometimes foreign to medical professionals, who see patients as either entering into, or not entering into, an arena that professionals control. The societal authority of medicine as an institution and the interpersonal authority of patients as choosers too often come to impasse, and individuals, voting with their feet, decline to enter the arena of medicine at all.

Communities define this problem as one of shaping more responsive medical institutions; thus they find themselves struggling to answer the necessary first question: What *is* an appropriate response? With this question we have returned to the root of the problem – if we can name our goals we can begin to achieve them, yet we seem to have too many unclear, unnamed, contradictory goals.

One other thing we now know is that none who has so far tried to wear the captain's hat can wear it alone, or ought to. Shared authority is the only viable possibility. There are many ways of sharing, however, and few of them work well.

Between patients and physicians, "shared decisionmaking" has been suggested as an appropriate sharing of authority about treatment choice. The complexities of the relationships we have elucidated in this volume alone suggest that this is too simple and slippery a solution. Patients and physicians are both authorities regarding different bodies of knowledge relevant to the choice to be made; and patient and physicians both wield different powers of absolute command: only the physician can give orders and perform procedures, and only the patient can, finally, say yes or no. Sharing in modern medicine can only mean cooperation, consultation, collaboration that takes account of complementary skill and knowledge and respects complementary powers of action and control; not a 50–50 split but something that is more formal and yet is also more fluent.

The primary focus of this volume represents a still greater challenge, however. That is the attempt similarly to articulate a model of shared authority in the societal realm. Not just two sharers but many have strong claims to power here. The same cautions against oversimple solutions obtain with even greater force; we cannot simply add up all the claimants to authority or all the goals of society and vote on them. Sharing here requires many things: public discussion of the goals of health care and of the skill, virtue, and knowledge needed to achieve them; identification of the role and scope of governmental and private action in service of those goals; finding the range of balances possible between health care and other social goals; naming the values that further our goals and social structures and are furthered by them.

What saves this formidable task from collapsing under its own weight is the strength of our shared desire for the clarity it demands: moral clarity, clarity of speech, clear knowledge, clear choices. Our task is already being tackled from all sides, as the essays in this volume have shown; we have simply to take care that each new insight is also turned toward helping us see the whole.

We may sail on for some time before we become able to control our course. But meanwhile things are coming slowly clearer – clear enough to show us, if not where we are going, then where we are. Since we are all in this boat together, that seems as good a place as any to begin.

NANCY M. P. KING

NOTES ON CONTRIBUTORS

Judith C. Areen, J.D., is Professor of Law and Associate Dean, Georgetown University Law Center, and Professor of Community and Family Medicine, Georgetown University Medical School, Washington, DC.

David Barnard, Ph.D., is Associate Professor, Department of Humanities, Milton S. Hershey Medical Center, Pennsylvania State University, Hershey, PA.

Larry R. Churchill, Ph.D., is Associate Professor, Departments of Social and Administrative Medicine and Religious Studies, School of Medicine, University of North Carolina, Chapel Hill, NC.

Robert M. Cook-Deegan, M.D., is Senior Analyst and Project Director, Office of Technology Assessment, Congress of the United States, and Research Fellow, Kennedy Institute of Ethics, Georgetown University, Washington, DC.

Alan W. Cross, M.D., is Associate Professor, Departments of Social and Administrative Medicine and Pediatrics, School of Medicine, University of North Carolina, Chapel Hill, NC.

Raymond S. Duff, M.D., is Professor of Pediatrics, Yale University School of Medicine, New Haven, CT.

H. Tristram Engelhardt, Jr., Ph.D., M.D., is Professor, Center for Ethics, Medicine and Public Issues, Baylor College of Medicine, Houston, Texas.

Nancy M. P. King, J.D., is Assistant Professor, Department of Social and Administrative Medicine, School of Medicine, University of North Carolina, Chapel Hill, NC.

Ernest N. Kraybill, M.D., is Professor of Pediatrics, School of Medicine, University of North Carolina, Chapel Hill, NC.

Joan E. Lynaugh, Ph.D., FAAN, is Associate Professor, School of Nursing, University of Pennsylvania, Philadelphia, PA.

Wendy K. Mariner, J.D., L.L.M., M.P.H., is Associate Professor of "Public Health (Health-Law)" and Socio-Medical Sciences and Community Medicine, Boston University Schools of Medicine and Public Health, and Lecturer, Social Medicine and Health Policy, Harvard Medical School, Boston, MA.

Russell C. Maulitz, M.D., Ph.D., is Director of Medical Student Programs, Presbyterian-University of Pennsylvania Medical Center, and

Lecturer in History and Sociology of Science in Medicine, University of Pennsylvania School of Medicine, Philadelphia, PA.

Peter J. Morris, M.D., M.P.H., is Associate Director, Wake County Department of Health, Raleigh, NC.

Stuart F. Spicker, Ph.D., is Professor, Department of Community Medicine and Health Care, University of Connecticut School of Medicine, Farmington, CT.

INDEX

The Philosophy and Medicine Book Series

Editors

H. Tristram Engelhardt, Jr. and Stuart F. Spicker

1. **Evaluation and Explanation in the Biomedical Sciences**
   1975, vi + 240 pp. ISBN 90–277–0553–4
2. **Philosophical Dimensions of the Neuro-Medical Sciences**
   1976, vi + 274 pp. ISBN 90–277–0672–7
3. **Philosophical Medical Ethics: Its Nature and Significance**
   1977, vi + 252 pp. ISBN 90–277–0772–3
4. **Mental Health: Philosophical Perspectives**
   1978, xxii + 302 pp. ISBN 90–277–0828–2
5. **Mental Illness: Law and Public Policy**
   1980, xvii + 254 pp. ISBN 90–277–1057–0
6. **Clinical Judgment: A Critical Appraisal**
   1979, xxvi + 278 pp. ISBN 90–277–0952–1
7. **Organism, Medicine, and Metaphysics**
   Essays in Honor of Hans Jonas on his 75th Birthday, May 10, 1978
   1978, xxvii + 330 pp. ISBN 90–277–0823–1
8. **Justice and Health Care**
   1981, xiv + 238 pp. ISBN 90–277–1207–7 (HB)/90–277–1251–4 (PB)
9. **The Law-Medicine Relation: A Philosophical Exploration**
   1981, xxx + 292 pp. ISBN 90–277–1217–4
10. **New Knowledge in the Biomedical Sciences**
    1982, xviii + 244 pp. ISBN 90–277–1319–7
11. **Beneficence and Health Care**
    1982, xvi + 264 pp. ISBN 90–277–1377–4
12. **Responsibility in Health Care**
    1982, xxiii + 285 pp. ISBN 90–277–1417–7
13. **Abortion and the Status of the Fetus**
    1983, xxxii + 349 pp. ISBN 90–277–1493–2
14. **The Clinical Encounter**
    1983, xvi + 309 pp. ISBN 90–277–1593–9
15. **Ethics and Mental Retardation**
    1984, xvi + 254 pp. ISBN 90–277–1630–7
16. **Health, Disease, and Causal Explanations in Medicine**
    1984, xxx + 250 pp. ISBN 90–277–1660–9
17. **Virtue and Medicine**
    Explorations in the Character of Medicine
    1985, xx + 363 pp. ISBN 90–277–1808–3
18. **Medical Ethics in Antiquity**
    Philosophical Perspectives on Abortion and Euthanasia
    1985, xxvi + 242 pp. ISBN 90–277–1825–3 (HB)/90–277–1915–2 (PB)
19. **Ethics and Critical Care Medicine**
    1985, xxii + 236 pp. ISBN 90–277–1820–2
20. **Theology and Bioethics**
    Exploring the Foundations and Frontiers
    1985, xxiv + 314 pp. ISBN 90–277–1857–1

21. **The Price of Health**
1986, xxx + 280 pp. ISBN 90–277–2285–4
22. **Sexuality and Medicine**
Volume I: Conceptual Roots
1987, xxxii + 271 pp. ISBN 90–277–2290–0 (HB)/90–277–2386–9 (PB)
23. **Sexuality and Medicine**
Volume II: Ethical Viewpoints in Transition
1987, xxxii + 279 pp. ISBN 1–55608–013–1 (HB)/1–55608–016–6(PB)
24. **Euthanasia and the Newborn**
Conflicts Regarding Saving Lives
1987, xxx + 313 pp. ISBN 90–277–3299–4 (HB)/1–55608–039–5(PB)
25. **Ethical Dimensions of Geriatric Care Value Conflicts for the 21st Century**
1987, xxxiv–298 pp. ISBN 90–55608–027–1
26. **On the Nature of Health**
An Action-Theoretic Approach
1987, xviii + 204 pp. ISBN 1–55608–032–8
27. **The Contraceptive Ethos**
Reproductive Rights and Responsibilities
1987, xxiv + 254 pp. ISBN 1–55608–035–2
28. **The Use of Human Beings in Research**
With Special Reference to Clinical Trials
1988, ISBN 1–55608–043–3
29. **The Physician as Captain of the Ship**
A Critical Appraisal
1988, ISBN 1–55608–044–1
30. **Health Care Systems**
Moral Conflicts in European and American Public Policy
1988, ISBN 1–55608–045X
31. **Death: Beyond Whole-Brain Criteria**
1988, ISBN 1–55608–053–0

GPSR Compliance

The European Union's (EU) General Product Safety Regulation (GPSR) is a set of rules that requires consumer products to be safe and our obligations to ensure this.

If you have any concerns about our products, you can contact us on ProductSafety@springernature.com

In case Publisher is established outside the EU, the EU authorized representative is:

Springer Nature Customer Service Center GmbH
Europaplatz 3
69115 Heidelberg, Germany

Batch number: 10761588

Printed by Printforce, the Netherlands